GAS! GAS! QUICK, BOYS!

GAS! GAS! QUICK, BOYS!

HOW CHEMISTRY CHANGED THE FIRST WORLD WAR

MICHAEL FREEMANTLE

For Martha and Ariane

First published 2012
by Spellmount, an imprint of

The History Press
The Mill, Brimscombe Port
Stroud, Gloucestershire, GL5 2QG
www.thehistorypress.co.uk

British Library Cataloguing in Publication Data.
A catalogue record for this book is available from the British Library.

ISBN 978 0 7524 6601 9

Typesetting and origination by The History Press
Printed in Great Britain

Contents

	Introduction	7
1.	The Chemists' War	11
2.	Shell Chemistry	27
3.	Mills Bombs and other Grenades	51
4.	The Highs and Lows of Explosives	59
5.	The Metals of War	83
6.	Gas! GAS! Quick, boys!	115
7.	Dye or Die	151
8.	Caring for the Wounded	159
9.	Fighting Infection	169
10.	Killing the Pain	199
11.	The Double-edged Sword	211
	Notes	215
	Bibliography	232
	Index	236

Introduction

My uncle, George Curtis, fought in the First World War and survived. Sadly, he died in 1966 and I never thought to ask him about his experiences in the conflict. My grandfather, Samuel, on my father's side may also have fought in the war, but once again it did not occur to me to ask him or my parents whether he did or not. He died soon after my uncle and my parents are no longer alive.

Until 2009, my interest in history had mainly been confined to the history of science. Indeed, I had written about various aspects of the subject over the preceding twenty-five years or so. As far as the First World War was concerned, I had read books such as Siegfried Sassoon's *Memoirs of an Infantry Officer*, watched films like *Oh! What a Lovely War*, and was familiar with Wilfred Owen's poem 'Dulce et Decorum Est', but that was the extent of my interest.

Then, in July 2009, my wife Mary and I travelled to Belgium and northern France to visit some of the First World War battlefields and cemeteries. For part of the trip we were joined by our son Dominic, his wife Claire, and their baby daughter Eloise. Both Mary and Dominic had been passionately interested in the war for many years.

One morning during our visit to the battlefield sites around Ypres, including Essex Farm, Passchendaele, Sanctuary Wood, and Langemarck (where the Germans launched their first chlorine gas assault), our tour guide informed us that the chlorine not only gassed the Allied troops but also dissolved in the muddy water that filled the shell holes to form highly corrosive hydrochloric acid. I was puzzled as I seemed to recall that chlorine dissolved in water to form chlorine water, a weakly acidic solution. Furthermore, chlorine-releasing chemicals were and still are used to purify drinking water and sterilise swimming pools.

Our guide then explained that the troops initially attempted to protect themselves from the gas by covering their noses and mouths with handkerchiefs or other cloths soaked in water or urine, adding that buckets of chemicals were subsequently provided for this purpose. As I have spent all my professional life working in the chemical industry, teaching chemistry, and writing about the subject, I was naturally interested in what chemicals were used, but the guide did not know.

This was to spark my interest in the chemistry of the First World War and, very soon, I was hunting for answers to questions like: What is lyddite? What is guncotton? What is ammonal? What is gunmetal? What is mustard gas? I found that the answers are out there in the public domain, but that they are scattered around in various books, articles, reports, museums and websites. What I could not find was a single volume that brought together all the different aspects of the chemistry of the war. That was surprising because chemistry, as I was soon to realise, was the *sine qua non* of the war. Indeed, some people called it 'the chemists' war'.

I therefore decided to write such a book and, specifically, a book for the general reader and not just for the many chemists who are interested in the First World War. The pages that follow describe the explosives, chemical warfare agents, metals, dyes, and medicines that were used in the war. They also show how chemistry and chemicals not only underpinned the war but also changed the war. Undoubtedly, without the advances in chemistry the war would have been much shorter and the death toll substantially reduced. Finally, the book reveals how much of the chemistry of the war evolved from discoveries and inventions made in the hundred or so years that preceded the war.

The majority of people who contributed information and comments for this book are no longer alive. They are the chemists and their fellow scientists, engineers and industrialists who described the chemistry of the war in reports, articles and books published either during the war or soon after. And they are the nurses and medical officers who cared for the sick and wounded, and also published accounts of their wartime experiences.

In particular, I have relied heavily on the *Journal of Industrial and Engineering Chemistry*, published monthly by the American Chemical Society until 1922, and the *British Medical Journal*, a weekly publication. I scoured every issue of both journals published during and immediately after the war for information and comments that I could use in this book – what appears is only the tip of the iceberg.

My visits to the Imperial War Museum in London, the In Flanders Fields Museum in Ypres, and the Somme Trench Museum in Albert have also yielded useful insights. In addition, I had some fascinating discussions about the war with Ken Seddon, a chemistry professor at Queen's University, Belfast and an expert on chemical warfare in general and phosgene in particular. Geoffrey Rayner-Canham, chemistry professor at Memorial University of Newfoundland, and his co-researcher Marelene Rayner-Canham, provided me with information about British female chemists and the First World War. Paul Gallagher, media relations executive at the Royal Society of Chemistry, alerted me to the discovery of the chemist who developed dyes for army and navy uniforms but was lost on the *Titanic*. Similarly, Catherine Duckworth, community history manager at Accrington Community History Library, Lancashire helped me considerably with the section on Frederick Gatty who patented and made a fortune out of mineral khaki dye. My thanks to them all.

I am also grateful to Jo de Vries, Paul Baillie-Lane and their colleagues at The History Press for commissioning this book and their work on its production; and to my wife Mary, our son Dominic and our three daughters, Helen, Charlotte and Lizzie, all of whom made valuable suggestions about the nature, content and not least the title of this book.

Michael Freemantle
Basingstoke
March 2012

Michael Freemantle is a science writer. He was Information Officer for IUPAC (International Union of Pure & Applied Chemistry) from 1985 to 1994. His duties included editing the IUPAC news magazine *Chemistry International*. From 1994 to 2007 he was European Science Editor/Senior Correspondent for *Chemical & Engineering News* – the weekly news magazine of the American Chemical Society. He was then appointed Science Writer in Residence at Queen's University, Belfast and Queen's University Ionic Liquid Laboratories for three years until 2010. Freemantle has written numerous news reports and articles on chemistry, the history of chemistry, and related topics. He is the author, co-author, or editor of some ten books on chemistry, including the textbooks *Essential Science: Chemistry* (Oxford University Press, 1983) and *Chemistry in Action* (Macmillan, 1987). His previous book, *An Introduction to Ionic Liquids*, was published by RSC (Royal Society of Chemistry) in November 2009.

I have been on battlefields; I know what war means. I have been in hospitals – I go to them yet – filled with the wreckage of this war; and when you think this is but a sample, and count up the cost of this war – nine or ten million men in the flower of their youth, the strongest, the most virile out of all the most civilised nations on the earth, *dead*; when you count the orphanage and widowhood, the withdrawal of that vast energy from the productive forces of civilization; when you think of the waste of only the material side, the amount they spent in money, two hundred thousand million dollars – and if you try to get some idea of what that is, you look in the world almanac and find the total value of the United States – of all the real and personal property in it, all the houses and all the lands and all improvements thereon since they took it from the Indians, telegraphs, jewelry and money, all of it added together amounts to one hundred and eighty-six thousand million dollars – when we think of these things we are filled with amazement. We have paid, not with king's ransom, but with the price of civilization and we have wasted a heritage greater in value than the aggregate value of the greatest country that ever existed on the face of the earth.

Newton D. Baker, U.S. Secretary of War, in an address on 'Chemistry in Warfare,' presented at the 58th meeting of the American Chemical Society, Philadelphia, USA, 3 September 1919. (*Ind. Eng. Chem.*, 1919, 11, p.921)

1

The Chemists' War

Applied chemistry, to a large extent

> Modern war, whether it be for robbing, plundering and subjugating other nations, or for legitimate self-defence, has become primarily dependent upon exact knowledge, good scientific engineering and, to a large extent, applied chemistry.

Few people outside the world of chemistry and industry will be familiar with the Belgian-American industrial chemist who made this remark at a meeting in New York on 10 December 1915. But many people will have heard of the plastic he invented and patented in 1907. The chemist was Leo Baekeland (1863–1944) and the plastic Bakelite.

The modern war to which Baekeland referred was the First World War, although, at the time, the United States had yet to enter the war and it was referred to by many Americans as 'the European War'. The war, also known as the 'Great War' and the 'War to End All Wars', was fought between the Allied Powers and the Central Powers from 1914 to 1918. The Allied Powers consisted of Britain, France, Japan, Russia and Serbia, with Italy joining in 1915, Portugal and Romania in 1916, and Greece and the United States in 1917. The Central Powers consisted of Germany, the Austro-Hungarian Empire and Ottoman Turkey, with Bulgaria joining them in 1915.

Baekeland's comment linking modern war to applied chemistry appeared as the opening paragraph of the published version of an address entitled 'The Naval Consulting Board of the United States', which he presented to members of the American Chemical Society and the American Electrochemical Society.[1] Baekeland, who was a member of the board, did not mention Bakelite or other early plastics in his address but focused rather on ways that the United States could harness its scientific and technical potential for the development and production of munitions and military equipment needed for defence.

Baekeland's audience in New York would have regarded his observations that modern war relied on applied chemistry as self-evident. They would have also known that chemistry and chemicals were not only being employed to devastating effect in the First World War, but also that chemistry in one form or another had been applied wittingly or unwittingly to warfare since time immemorial.

Over 2,000 years ago, for example, Roman legionaries in battle wore armour made of iron, a chemical element, and helmets made of bronze, an alloy – a mixture of a metal and one or more other chemical elements. The composition of bronzes varies but is typically 80 per cent copper and 20 per cent tin, and, like iron, these two metallic elements are extracted from ores by smelting. The operation is based on two chemical processes: first, the ores are converted to metal oxides; then the oxygen is removed from the oxides by heating them with what is known as a reducing agent, typically the chemical element carbon in the form of charcoal or coke.

Gunpowder provides another example of the application of chemistry to warfare. The powder consists of a mixture of charcoal, the chemical element sulfur and one chemical compound – potassium nitrate. Its use in warfare dates back to the introduction of the gun as a weapon in the fourteenth and fifteenth centuries. In fact, gunpowder chemistry also played a role in the birth of modern chemistry as we now know it. The birth is widely attributed to the publication of the first chemistry textbook *Traité Élémentaire de Chimie* (Elementary Treatise on Chemistry) by French chemist Antoine Laurent Lavoisier (1743–1794) in 1789. From 1776 to 1791, Lavoisier was responsible for gunpowder production and research at the Royal Arsenal in France. Unfortunately for him, he was also a tax collector and served on several aristocratic administrative councils. During France's 'Reign of Terror', he was accused by revolutionists of counter-revolutionary activity, found guilty and executed by guillotine on 8 May 1794.

In the following century, chemistry as a distinct discipline and profession began to take root. For example, the Chemical Society of London, the French Chemical Society and the American Chemical Society were founded in 1841, 1857 and 1876 respectively, while the Society of Chemical Industry was formed in Britain in 1881 and its German equivalent in 1887. The first ever international meeting of chemists took place at a congress in Karlsruhe, Germany, in 1860 and the participants included the Russian chemists Dmitri Mendeleev (1834–1907), the chief architect of the Periodic Table of Chemical Elements, and Alexander Borodin (1833–1887), a respected chemist who is best remembered as a composer. Another international conference on chemistry took place in Paris in 1889 and the first of a series of International Congresses on Applied Chemistry was held in Brussels in 1894.

The application of chemistry to warfare also developed rapidly in the years between Lavoisier's 1789 treatise and the outbreak of the First World War. The discovery of new types of powerful explosives, new medicines and drugs to treat wounded soldiers, and new types of metal alloys for weapons and military equipment during this period had a major impact on the war.

One of the discoveries in applied chemistry during the nineteenth century which, perhaps surprisingly, had immense significance for the First World War was the synthesis of mauve, the first synthetic dye, by English chemist William Henry Perkin (1838–1907) in 1856. Other synthetic dyes soon followed; for example, two years after the discovery of mauve, Perkin's chemistry teacher, the German chemist August Wilhelm von Hofmann (1818–1892) synthesised the dye magenta. These dyes, like their natural counterparts, are organic compounds – they contain the element carbon.

The discoveries of synthetic dyes not only revolutionised fashion and the textiles industry, but they also gave birth to the synthetic dyes industry and the mass production of organic chemicals. Furthermore, they sparked widespread interest in the commercial applications of synthetic organic chemistry. Before then, the organic chemical industry had been largely confined to the manufacture of soap from fats and oils.

Germany was the quickest to recognise the commercial potential of synthetic organic chemistry. In the years leading up to the First World War, the country became the world's predominant manufacturer and exporter of synthetic dyes and other commercially-important, synthetic, organic compounds, most notably pharmaceutical products. Furthermore, soon after the beginning of the First World War, Germany was able to adapt the chemical plants in its dye-producing factories for the industrial-scale production of trinitrotoluene (TNT) and other powerful explosives based on organic compounds.

The discovery and development of synthetic plastics also occurred in the nineteenth century. In the 1860s, English chemist Alexander Parkes (1813–1890) invented celluloid, the first synthetic plastic. American inventor John Wesley Hyatt (1837–1920) subsequently developed the synthesis for a variety of commercial applications and it was used to make the photographic roll film developed by George Eastman (1854–1932) in 1889. By the start of the First World War, photography had become sufficiently advanced not only to record life and death on the frontline, but also for training and reconnaissance purposes.

Richard B. Pilcher, Registrar and Secretary of the Britain's Institute of Chemistry, refers to several examples of chemistry's applications to the war effort in an article published in the journal in September 1917.[2] He notes that professional chemists could provide 'efficient service in the many requirements of the naval, military, and air forces.' He explains, for instance, that the service of chemists was essential to control the manufacture of munitions, explosives, metals, leather, rubber, oils, gases, food and drugs. His list does not include, but might well have included, the manufacture of antiseptics, disinfectants, anaesthetics, synthetic dyes and photographic materials. Chemists, he continues, were also needed for the analysis of all these materials as well as for the analysis of water: the detection of poisons in streams, the disposal of sewage and other matters of hygiene.

Pilcher calls the First World War the 'Chemists' War', and the term has since been used by many others for the war, including David J. Rhees, executive director of the Bakken Library and Museum in Minneapolis, USA, who wrote in an article on the war published in the early 1990s:

> When we speak of World War I as the Chemists' War, the image that usually comes to mind is the famous battle near the Belgian town of Ypres where, on 22 April 1915, the Germany army released a greenish-yellow cloud of chlorine gas on Allied troops.[3]

In many people's eyes, the use of chemicals in the First World War has become synonymous with chemical warfare and the use of poison gases against enemy troops. The active and creative role played by chemists in this type of warfare inevitably contributed to the subsequent widespread negative image of chemicals: 'To call the Great War a "Chemists' War" was perhaps a matter of pride, but not exactly for praise', remarks Roy MacLeod, an historian at the University of Sydney, Australia, in an article published in 1993.[4]

The term has a much broader context, however. The chemistry of the First World War was not just confined to poison gases and explosives, but also to the development and production of numerous other chemical products used by the military either directly or indirectly: 'Many regard the war as largely a conflict between the men of science of the countries engaged', observes Pilcher, implying that the side that mastered the chemistry needed for warfare would be successful in the war. Germany had an advantage in this respect in the years leading up to the war, especially in the number of professional chemists who could contribute to the war effort. According to historian Michael Sanderson, in 1906 an estimated 500 chemists worked in the British chemical industry whereas there were 4,500 in the German chemical industry.[5] By the start of the war, there was one university-trained chemist for every fifteen workers engaged in the German chemical industry and one for every forty workers in German industry as a whole.[6] In Britain, on the other hand, there was one university-trained chemist for every 500 workers employed in the various industries.

In London, education, training, and research in key areas of chemistry were also lacking. Although the subject was taught well, there was little or no emphasis on applied chemistry and industrial chemistry. 'Most scandalous', Sanderson comments, was the 'notoriously casual' attitude to the use of coal tar. Coal tar, produced as a by-product when coal is converted into coke or coal gas, was an important source of organic chemicals used for the manufacture of dyes, pharmaceuticals, explosives and other products. Before the war, coal tar produced in Britain was exported as a raw material to Germany, and even though the first coal tar dye, mauve, had been synthesised by Perkin, an English chemist, it was German industry that exploited the expertise of its chemists to attain virtually a world monopoly in the manufacture of chemical products derived from coal tar. Following Perkin's discovery, the organic chemical industry in Britain rapidly declined and did not recover until after the First World War.

The story was similar in other Allied countries such as France. In 1885, for example, French chemist François Eugène Turpin (1848–1927) patented the use of the pressed or fused form of the organic chemical picric acid as a fragmentation charge for artillery shells. The French government subsequently adopted the explosive for its high-explosive shells. The explosive was known in France as melinite whereas in Britain it was called lyddite. Picric acid, or trinitrophenol to give it its full chemical name, was made from phenol, a coal tar chemical, and nitric acid. Yet in 1914, French supplies of phenol for manufacture of the explosive came from foreign countries and particularly from Germany.[7] Furthermore, prior to the war, France manufactured relatively few coal tar-based pharmaceuticals and instead relied extensively on imports of pharmaceuticals from Germany.[8]

There was also one other major issue that not only influenced the duration of the First World War, but also demonstrated the professional expertise of German chemists. That issue was nitrogen, a chemical element that comprises roughly 80 per cent of the air around us.

Germany's nitrogen problem

At the beginning of the war in August 1914, there was widespread belief that the conflict would be short and almost certainly over within a few months. Germany entered the war with stocks of ammunition for an intensive campaign of just a few months. However, with the onset of trench warfare in September 1914, it soon became apparent that the progress of the war would be slow and Germany's stocks of ammunition rapidly diminished. The country therefore mobilised its national industries, including the chemical industry, to restock its stores to prepare for a longer campaign.

The German chemical industry had to adapt rapidly. One of its major problems was the manufacture of nitric acid which was needed to make explosives such as TNT and picric acid. Both of these explosives are nitrogen-containing organic chemicals, with coal tar the source of both the carbon and nitrogen for these chemicals. However, nitrogen was also needed to manufacture fertilisers and the limited supplies of the type of coal suitable for the production of the coal tar reduced the amount of nitrogen-containing chemical compounds that could be produced. Germany therefore used a nitrate mineral as a supplementary source of nitrogen.

Until the outbreak of war, Germany imported the mineral from Chile. It was then converted to nitric acid by reaction with sulfuric acid. However, the British naval blockade cut off nitrate supplies and Germany found itself with insufficient nitrogen to manufacture its fertilisers and explosives. The German chemical industry therefore turned to 'nitrogen fixation', a process that converts nitrogen in the air into nitrogen-containing compounds.

The specific nitrogen fixation process used by the German industry was based on a discovery in 1908 by German chemist Fritz Haber (1868–1934). He showed that ammonia, a compound containing the elements nitrogen and hydrogen, could be synthesised by the reaction of the two elements in their gaseous forms in the presence of iron. The iron functioned as a catalyst, increasing the rate of the reaction in a process known as catalysis. In 1918, Haber won the Nobel Prize in Chemistry for the discovery.

Carl Bosch (1874–1940), an industrial chemist working for the German chemical firm BASF, subsequently designed a reactor that allowed the Haber process to be carried out at high pressures and temperatures. The company started producing ammonia using the Haber-Bosch process, as it became known, in 1913. Bosch also won the Nobel Prize in Chemistry in 1931 for his contribution 'to the invention and development of chemical high pressure methods'.

At first, the ammonia produced by this process was used to make ammonium sulfate, a soil fertiliser. However, it was well known that ammonia could also be converted to nitric acid using a method developed by another German chemist, Friedrich Wilhelm Ostwald (1853–1932). The process combines ammonia with atmospheric oxygen in the presence of a platinum catalyst to form a nitrogen- and oxygen-containing gas called nitrogen monoxide. The gas is then oxidised with atmospheric oxygen to yield a related gas, nitrogen dioxide. Nitric acid is produced by passing the nitrogen dioxide gas through water. Ostwald patented the process in 1902 and won the Nobel Prize in Chemistry in 1909 in recognition of his work on catalysis and also for other chemistry research he had undertaken.

In his New York address in 1915, Baekeland observed that Germany would have been 'hopelessly paralysed' had it not been for the development of chemical processes for the manufacture of nitric acid from air. If these processes, developed by German chemists in the early twentieth century, had not been available to German industry, the war may well have come to a conclusion by the end of 1914 – as had been widely predicted at the beginning of the war.

The importance of electrochemistry

'Never before in the history of electrochemistry has the vast importance of the various electrochemical products been so forcibly brought to the attention of our government and of our people as the present year of the Great War', remarked

Colin G. Fink, President of the American Electrochemical Society, in September 1917.[9] He was speaking at Third National Exposition of Chemical Industries which was held in New York, just a few months after the United States had declared war on Germany.

Electrochemistry focuses on the electronic aspects of chemistry and the relationship between electricity and chemistry. It is concerned with the impact of electricity on chemicals and, conversely, on the use of chemicals to generate electricity. This branch of chemistry is an important component of the science and technology of metals and alloys, otherwise known as metallurgy. The armour, artillery, munitions, tanks, aircraft, battleships and, of course, railways of the First World War all relied on expertise in metallurgy and electrochemistry for their manufacture and construction. 'Take from this country its electrochemical industry with its numerous and diversified manufactures and the martial strength of our country is hopelessly crippled', Fink said, pointing out, for example, that thousands of rifles and guns were turned out every month with the steels made by the electric arc furnace.

All steels are made of iron and a small percentage of carbon. Mild steel, which contains just 0.2 per cent carbon, is malleable and ductile, and was used in the Great War to make barbed wire and other products. The hardness of carbon steels, as they are called, is increased by increasing carbon content. Steels, known as alloy steels, contain not only iron and small amounts of carbon but also up to 50 per cent of one or more other metallic elements, such as aluminium, chromium, cobalt, molybdenum, nickel, titanium, tungsten and vanadium. The addition of these metals improves the properties of the steels. Tungsten, for example, improves the hardness, toughness and heat resistance of steel.

The development of the electric arc furnace in the 1890s and early twentieth century added a new dimension to steel manufacture. When the electric power of the furnace is switched on, temperatures are generated that are sufficiently high to melt scrap iron and steel, which enabled it to be converted into the high-quality alloy steels needed for the war effort. Numerous alloys, produced by the electric furnace, were used in nearly every item of the United States government's vast military equipment for the war.

William S. Culbertson, in another speech at the New York exposition, agreed with Fink on the importance of the electric arc furnace in revolutionising steel manufacture.[10] He describes silicon steel, for example, as 'indispensable' in the manufacture of munitions, adding that steels containing the metallic element tungsten improved the efficiency of metal cutting tools, while the addition of chromium, nickel, vanadium or molybdenum conferred special properties to steel, 'making it peculiarly suited to many special uses, including armour plate'.

Electrolysis also played a key role in producing the metals and a range of other chemicals required by the military during the war. In this electrochemical technique, chemical reactions take place when an electric current is passed through an electrolyte contained in an electrolytic cell. The electrolyte is typically a molten salt or a solution of a salt in water.

After extraction from its ores, pure copper was produced by electrolysis in a process known as electrorefining. In his speech, Fink pointed out that large quantities of 'electrolytic copper' were 'absolutely essential' for the manufacture of electrical apparatus, as were sufficient quantities of aluminium and magnesium, two metallic elements that were also produced using electrolysis and used for the light, strong stays of aircraft. Similarly, Fink added that liquid chlorine, which was used to synthesise some of the chemicals in preparations for 'treating the wounds of our heroes', was a product of electrolysis, as was hydrogen used 'in all of our scout and observation balloons'. He continued:

> May we continue to lead the world in the supply of the many electrochemical products, pure metals and alloys for the arts, gases for cutting and welding, chlorine and peroxides for our hospitals, chlorates and acetone for munitions, nitrates for the farm and defense, abrasives, electrodes, solvents and lubricants! May we continue to excel in the products of the electric furnace and the electrolytic cell!

Fink failed to mention that the chlorine generated in electrolytic cells was used as a poison gas by both sides during the Great War. It was also used to synthesise lethal chlorine-containing gases such as phosgene that were employed to devastating effect against entrenched enemy troops.

In September 1918, F. J. Tone, President of the American Electrochemical Society, speaking in New York at the Fourth National Exposition of Chemical Industries, provided a graphic example of the importance of electrochemistry and metallurgy to the United States' aircraft programme.[11] He noted that the crank cases and pistons of the motors in aircraft were made of aluminium, and that the crank shafts and engine parts, which were subjected to the greatest strains, were all composed of chrome alloy steel: 'All of these parts are brought to mechanical perfection and made interchangeable by being finished to a fraction of a thousandth of an inch by means of the modern grinding wheel made from electric furnace abrasives.'

Calcium carbide, a chemical compound consisting of calcium and carbon, was also made in the electric arc furnace, and the acetylene that was made from it facilitated an ample supply of aeroplane dope. The dope, a compound known as cellulose acetate, was used to tighten the fabric, often linen, skins covering aircraft. Tone continued:

> When the aviator trains his machine gun on an enemy plane his firing is made effective by tracer bullets of magnesium or phosphorus. When our bombing planes begin to carry the war to Germany it will be with bombs perhaps of

> ammonium nitrate or picric acid or other high explosives all depending largely in their manufacture on electrochemical reagents.

Not just a chemists' war

In an article published in 2005, George Bailey, a senior lecturer at the University of Westminster, lists a number of key elements in the transformation of the British Expeditionary Force (BEF) from the colonial-style army of 1914 to the victorious continental-style armies of 1918.[12] These key elements consisted of: manpower, volunteers and conscription; high command and the British War Cabinet; infantry weaponry and tactics; artillery; suppression of enemy infantry and artillery; tanks; aircraft; moving supplies; rescuing the wounded; and patriotism and maintenance of discipline.

Several of these key elements, such as artillery, tanks, and aircraft, relied on technology and the productive capacity of British industry. In many ways, the First World War could therefore be considered to be a war of technology or a war of industrial power in which scientists, engineers and factory workers in the opposing nations strove to produce superior military equipment and munitions in ever increasing quantities. At the same time, it was also necessary to sustain agricultural productivity in order to feed the people at home and the troops at the front; to transport troops, food and military equipment to the front and the troops back from the front; and to provide and improve medical supplies. All these activities relied to a greater or lesser extent on science, technology and engineering.

The war was a 'war of engineers' said David Lloyd George (1863–1945), who was Britain's Minister of Munitions from May 1915 to July 1916 and later became the country's prime minister. In an interview with a correspondent of *The New York Times* in November 1915, Lloyd George's Director of Recruiting for Munitions Work, Lord Murray of Elibank (1870–1920), observed that the ablest engineers in Great Britain, working in many of the country's largest engineering establishments, were driving forward 'enormously successful engineering work'.[13] He remarked that ships, guns, armour, shells, rifles, and bullets are pouring out of factories in an unending stream, adding: 'The stream is destined to increase steadily every month'.

His forecast was to prove correct: by the beginning of 1917, the production of high explosives in England was sixty-two times what it was in 1915, as H.E. Howe noted in an address delivered at a meeting of the American Chemical Society in Kansas City on 12 April 1917:[14]

> British munitions factories are now making more heavy gun ammunition every 24 hours than they manufactured during the entire first year of the war ... The monthly output of heavy guns is more than six times what it was during the year 1915.

Similarly, between May 1915 and May 1916, the output of bombs increased 33-fold and the quantity of machine guns manufactured during the twelve months up to August 1916 was fourteen times greater than in the twelve months up to August 1915.

The increased productivity in munitions factories relied not only on engineers and the efforts of factory workers, but also on inputs from chemists and chemical engineers to develop increasingly efficient chemical processes for manufacturing explosives and other chemical products of military value. Thus, the 'war of engineers' was underpinned by applied chemistry.

In his 1993 article, MacLeod refers to the 1914–1918 conflict as a 'scientific war'.[4] He notes, for example, that physicists developed acoustic devices, geographers prepared artillery maps, geologists designed trench systems and tunnels, surgeons applied triage, psychologists discovered shell shock, and bacteriologists and entomologists attempted to halt the spread of infectious diseases. And once again, applied chemistry underpinned much of this 'scientific war'. For example, the inks used to print the geographer's artillery maps contained insoluble black, white or coloured chemical materials known as pigments, and chemical preparations were used as disinfectants throughout the war to halt the spread of infection.

War: the mother of invention?

'War has always been the mother of invention' observes Oxford University historian A.J.P. Taylor (1906–1990) in the preface of his book *The First World War: An Illustrated History*, adding also that: 'Historical photographs are among her children'.[15] Taylor points out that the Crimean War (1853–1856) in which Russia fought against Britain, France, Piedmont and Turkey, raised photography 'from its infancy' and his book is illustrated with numerous black and white photographs that show the battles of the Great War, life and death in the trenches, the aircraft and battleships, and the generals and politicians who played key roles in the conflict.

War, however, did not give birth to photography. The invention of photography dates back to the 1830s when several people began to use cameras and chemicals to produce photographic images. In 1839, for example, Frenchman Louis Daguerre (1789–1851) perfected a process for producing a permanent photographic image, known as the 'daguerreotype', on a silver-coated copper plate treated with iodine vapour. The image was then developed using mercury.

In his book on chemicals in war, published in 1937, Augustin M. Prentiss cites three 'outstanding' innovative technological developments of the First World War: 'the military airplane, the combat tank, and chemical warfare'.[16] He explains that each of these new instruments of war made its appearance on the battlefield at about the same time and each exerted an important influence in shaping the character of modern combat. The author, a lieutenant colonel in the United States Chemical Warfare Service, might have added other developments such as submarines, torpedoes and floating mines.

These new or infant technologies relied not only on sophisticated engineering but also on a body of scientific and technical knowledge and discovery that predated the First World War. Chlorine, for example, is a chemical element famously used as a poison gas in the early days of the war, but it was not new. The element was discovered by self-taught Swedish chemist Carl Wilhelm Scheele in 1774. Similarly, radiography, which was used widely in hospitals during the war, has roots that can be traced back to November 1895 when the German physicist Wilhelm Konrad Röntgen (1845–1923) discovered X-rays. The Haber-Bosch process for the synthesis of ammonia from atmospheric nitrogen, on which Germany relied for the manufacture of its explosives, was also developed before Germany declared war on Russia at the beginning of August 1914.

The manufacture of guns, armour, explosives, tanks, aircraft, battleships, submarines and torpedoes, as well as photography and radiography, all relied on the use of chemicals and the application of chemistry discovered before the war.

Invention in chemistry during the war, however, was limited. Chemists in academe were diverted from their normal research activities in university chemistry laboratories to war activities, either in military service or to improving the productivity of the chemical materials, such as explosives, needed for war. Chemists in the Chemical Warfare Service in the United States worked to devise new and more effective chemicals to disable and kill enemy troops, while other chemists worked on pharmaceutical and antiseptic preparations used to treat wounded and sick soldiers.

What innovation there was often focused on the development of new types of poison gas, better explosives and more efficient processes for the manufacture of the chemicals needed for the war effort. 'The necessity of increased production of munitions involving all types of chemical mixtures and compounds requires chemists in large numbers, not only to inspect and analyze the substances we are using, but to develop the necessary new ones', noted Rear Admiral Ralph Earle, Chief of the Bureau of Ordnance, US Navy, in an address on 'Chemistry and the Navy', presented at the 58th meeting of the American Chemical Society in Philadelphia on 3 September 1919.[17] The course, duration and outcome of the war not only depended on access to the raw materials, such as nitrates, and the ability of chemists and chemical engineers to convert these into useful military materials and equipment, it also depended on their ingenuity to adapt and develop existing chemical technology for the war effort.

But did the First World War give birth to any major discoveries in chemistry? According to one chronicle of science, three of the most significant events in chemistry during the period 1914–1918 were the synthesis of acetic acid (1914), the discovery that cobalt increases the strength of tungsten steel when the alloy is magnetised (1916), and the award of the Nobel Prize in Chemistry to Haber for his synthesis of ammonia from hydrogen and atmospheric nitrogen (1918).[18] The war, however, did not give birth to any of these events; indeed, it would be difficult to claim that the Great War was the mother of any important inventions or discoveries in chemistry. Rather, the war harnessed, nurtured and facilitated improvements to earlier inventions in chemistry and related sciences such as metallurgy.

Impact of war on the chemical industry

Whereas applied chemistry underpinned much of the technology deployed in the First World War, the war conversely had a direct impact on the ability of the chemical and its associated industries to provide for civilian needs. This impact was felt acutely in the United States at the beginning of the war.

Culbertson pointed out in his New York address in 1917 that daily life in the country depended on chemicals: textiles required dyes; chemicals were needed for refining sugar and petroleum and for the manufacture of glass, pottery, paper, paints, varnishes, rubber and cement; and medicinal and pharmaceutical products, toilet preparations, photographic materials, motion picture films, cleaning compounds, baking powder all relied on the chemical industries. The chemical industries also supplied fertilisers to agriculture and materials for the tanning industry.[10]

At the outbreak of war in August 1914, an editorial in *The Journal of Industrial and Engineering Chemistry* pronounced that 'the war in Europe will soon be over', and saw the war as a chance for the chemical industry in the United States to prosper:[19] 'While the eyes of the world are turned upon the military activities of Europe, business strategists in the United States will not fail to recognize the tempting opportunities for making ourselves more independent of foreign supplies.'

In 1914, the United States already had a strong chemical industry which supplied many of the country's needs. The country produced more sulfuric acid than any other in the world, and possibly more than all other countries combined, observed William. H. Nichols in an address on 'The War and the Chemical Industry' that he presented at a meeting of the American Association for the Advancement of Science in Philadelphia in December 1914.[20] Sulfuric acid is an important commodity chemical that is used in a wide range of applications, such as explosives, dyestuffs, fertilisers and in the manufacture of other acids. German organic chemist Justus von Liebig famously remarked in 1843: 'We may fairly judge the commercial prosperity of a country by the amount of sulfuric acid it consumes.'

In his address, Nichols pointed out that the United States had an abundance of many of the raw materials needed for the manufacture of chemical products, including cheap phosphate rock, salt, copper, coal, wood, bauxite and zinc. Yet many raw materials were also imported into the country, including sulfur in the form of the ore pyrites, potash, tin, nickel and sodium nitrate.

The 'European war' raised an important question for chemists and chemical engineers in the United States, as Edward Gudeman commented in an address at the annual meeting of the American Institute of Chemical Engineers, Philadelphia, December 1914: 'Will and can United States chemists prove equal to the emergency created, which makes it impossible, in many cases, to obtain from abroad, many chemical supplies for which the United States in the past has been and is today and will be in the future, dependent on foreign producers?'[21] The supply of synthetic organic chemicals and coal tar dyes which the United States had imported from Germany before the war was a major challenge. When the war broke out, the country was importing about 90 per cent of its coal tar dyes from Germany, and American dye companies manufactured the remaining 10 per cent from chemicals supplied by Germany. The British naval blockade cut off the supply of these German dyes and chemicals resulting in the so-called 'dye famine' in the United States.

Before August 1914, the United States was practically dependent on Germany for colour[10] and the onset of war resulted almost in a panic among those who used dyes: 'Many lines of our industrial life were threatened with utter demoralization because of the shortage of dyestuffs and medicinals resulting from the blockade of German ports by the British navy', notes an editorial in *The Journal of Industrial and Engineering Chemistry*.[22] 'Textile mills faced the imminent possibility of shutting down because of the inability to secure dyestuffs for their fabrics. Tanners, lithographers, and wallpaper men sought in vain for needed coloring matter and pharmacists' stocks of many much-used medicinals became depleted.'

Domestic manufacturers in the United States soon realised that the dye famine opened up a golden window of opportunity. A number of American companies, most of them small, urgently set out to manufacture dyes, pharmaceuticals and other chemical products derived from coal tar. Within three years of being cut off from the German supply, the country had invested huge sums in plants for manufacturing dyes and other organic chemicals:[10] 'We were producing as large a quantity of dyes as were consumed here when the war started', observed Culbertson. Baekeland also proudly pointed out that it had taken the United States less than three years to become independent of Germany in a line of chemicals, primarily dyes, which Germany had taken over fifty years to develop.[23]

Similarly, before the war, the United States had no synthetic phenol industry.[21] Phenol, a coal tar chemical, was used as a starting material to manufacture a wide variety of products for civilian and military use, including dyes, photographic developers, medicines, flavours, perfumes, and explosive materials such as picric acid. Within three years, however, the country had a thriving synthetic phenol industry, with chemist Grinnell Jones reporting in an article published in 1918 that fifteen chemical plants produced 64,146,499lb of phenol valued at $23,715,805 in 1917, most of which was used in making picric acid.[24]

The war inevitably increased demand not just for the home-produced chemical products needed for domestic use, but also for the chemicals needed for the war effort. The amount of the explosive TNT manufactured in the United States, for example, rose from 3.4 million pounds in 1913 to a rate of 16 million pounds per month in 1916, while a similar growth was reported in chlorine, potash, coal tar dyes and pharmaceuticals.[3]

The war thus kick-started important sectors of the chemical and its associated industries in the United States and exports of chemicals from the country rose markedly. W.S. Kies, Vice-President of the National City Bank of New York and American International Corporation, commented in an address on 'the development of our export trade' on 26 September 1917: 'The chemical industry of the United States has shown greater efficiency and greater powers of quick response to business demands than almost any other of the great industries of the country.'[25] Kies also pointed out that the value of exports of chemicals in 1917 (the fiscal year ending 30 June 1917) was $185 million, which was practically seven times the $27 million value for 1914. Exports of all industries, as a whole, were only three times as great. However, these export figures did not include explosives, Kies adding: 'In explosives the value of our exports grew from $6 million in 1914 to $820 million in 1917 … Under this class were listed cartridges, dynamite and gunpowder.' Exports of other types of explosives grew from $1 million in 1914 to $420 million in 1917, with these values showing that the growth of exports from industries allied to the chemical industries had been 'quite as striking' as from the chemical industries themselves. Kies summarised: 'During the last three years, the chemical industry has received a great impetus. Large amounts of money have been spent on its development. In many lines, before the war, Germany was supreme and competition with her was impossible. Germany, prior to the war, as we all know, had a grip on the chemical markets of the world.' By the end of the war, however, this was no longer the case as the naval blockade and the war had sparked a major boom in the American chemical and associated industries and its exports.

The final impact

In essence, the chemical and its associated industries use chemical processes, devised by chemists and chemical engineers, to convert raw materials – or intermediate chemicals obtained from these raw materials – into finished chemical products. The raw materials occur naturally in the land, water, air, and plant and animal life. These include carbon-containing raw materials such as coal, wood and petroleum, which eventually yield organic, that is carbon-containing, chemical products like dyes, explosives such as TNT, pharmaceuticals and plastics. Inorganic chemical products, such as steels, brasses and other alloys, fertilisers, nitric acid and other mineral acids, and numerous pigments are obtained from metal-containing ores, phosphorus-containing rocks, chlorine-containing salt, atmospheric nitrogen and other inorganic sources.

The First World War interrupted the supply of many of these raw materials and intermediate chemicals to the chemical and its associated industries in countries such as Germany and the United States. The war also caused the manufacture of chemical products to be redirected from normal civilian uses to military uses; the most obvious example being the diversion of chemical resources from agriculture to the military in Germany.

During the war, chemists were well aware of this problem, with Culbertson remarking in 1917: 'The factories that produce nitrogeneous fertiliser in time of peace will yield us nitric acid in time of war … Those producing intermediates and dyes can turn their machinery and workers to making explosives.'[10] The following year, Jones comments that the fertiliser industry had probably made the greatest sacrifices in this regard: 'Sodium nitrate and ammonia are required for the manufacture of explosives in such large quantities that the amounts left for use in fertilisers has been and will be much reduced.'[24]

In Germany, the diversion of raw materials and chemical products from civilian to military use was a significant factor in the country's defeat in 1918: 'Blockade, starvation and internal unrest had brought Germany to defeat' notes MacLeod in his 1993 article.[4] The blockade deprived Germany of wheat imports from the United States and beef imports from Argentina to feed the civilian population. It also cut off the supply of Chilean nitrates needed to make nitric acid and therefore forced the country to employ the Haber-Bosch process to convert atmospheric nitrogen to ammonia and the Ostwald process to convert the ammonia to nitric acid for the manufacture of explosives. The switch to explosives manufacture left insufficient capacity for the manufacture of the nitrogen-containing fertilisers needed to bolster food productivity and make up for the shortfall in imports. German agriculture and food production suffered as a consequence, leading to starvation, malnutrition and loss of morale on the home front in Germany.

MacLeod observes that Germany's science-based industry and its capacity to apply scientific methods was a 'formidable enemy' that almost saved the country from defeat in the First World War. Science-based industry could have been the single most decisive factor in an Allied victory had the war continued until 1919, although, MacLeod believes, the 'struggle might well have gone the other way'.

2

Shell Chemistry

Millions of shells

On 21 February 1916, Crown Prince Wilhelm's Fifth Army, led by Chief of Staff General Erich von Falkenhayn (1861–1922), launched an attack along an 8-mile front around Verdun. The city lies in Lorraine, a region in north-eastern France that borders Belgium, Luxembourg and Germany. The city was defended by the French Second Army and heavily fortified with a ring of forts.

The German Army stockpiled some 3 million shells for the battle, and during the first hour of the attack the German artillery bombarded the French forces with around 100,000 shells.[1] The battle continued until 18 December 1916. By July, according to one estimate, the French and Germans had fired a total of 23 million shells at each other, which works out at an average of around 100 shells per minute, day and night.[2] Shelling was even more intense during the Battle of the Somme (1 July–18 November 1916) when the British and Germany armies fired a total of 30 million shells at an average of almost 150 per minute. The battle resulted in some 419,000 British, 194,000 French and over half a million German casualties. The following year, the British artillery launched a massive artillery bombardment of German positions along the Messines Ridge, south of Ypres, prior to the Battle of Messines which was fought from 7 to 14 June. Over a ten-day period the artillery fired over 3.5 million shells, which equated to over three shells per second.

The intense shelling continued until the end of the war. On 21 March 1918, for example, the German Army launched an offensive, codenamed Operation Michael, against the British Third and Fifth armies who were holding a 50-mile section of the frontline in France between Arras and La Fère. On the first day of the battle, known as the 1918 Battle of the Somme or the Second Battle of the Somme, the Germans unleashed an artillery barrage of unprecedented ferocity: over a five-hour period their guns fire some 1 million shells.

Nothing symbolises the Great War more than shells and shelling, and they are a recurrent theme in the literature of the war. For example, in an early chapter of his memoir *Storm of Steel*, German soldier Ernst Jünger (1895–1998) portrayed life and death in the frontline at Les Eparges in north-eastern France in April 1915. Writing about coming under shell fire for the first time, he describes the 'savage pounding dance' of the shelling and the incessant flames from the artillery mingled with 'black, white and yellow clouds'.[3] He also describes how he received a flesh wound from 'a needle-sharp piece of shrapnel'. In *Goodbye to All That*, war poet Robert Graves (1895–1985) also recounts his experiences in the trenches during the war.[4] In May 1915 at Cuinchy in the Pas-de-Calais region of France, for example, one shell came 'whish-whishing' towards him, before he dropped flat in the trench and the shell burst just over the trench.

British troops called the heavy shells fired from German siege howitzers 'coal-boxes' because they generated large amounts of black smoke when they burst. And as the shells had such a powerful impact, they were also referred to as 'Jack Johnsons' after the African-American boxer Jack Johnson, who was world heavyweight boxing champion from 1908 to 1915.

Shell crises

By the early 1900s, Germany had become on industrial powerhouse on a global scale. The country exported machine tools, dyes, pharmaceuticals, and photographic and other chemical products to Britain, the United States and throughout the world. The Krupp company, based in Essen in the Ruhr area of Germany, was one of the country's largest and most important technology-based firms. It was an industrial empire which manufactured rails and wheels for railways, armaments and other high-quality steel products, many of which were exported.

David Lloyd George, who became Britain's Minister of Munitions in May 1915 observed that, at the start of the war, the Germans and Austrians between them had much larger supplies of war materiel and more extensive factories for turning out military supplies than the Allied countries. The chemical companies produced the explosives needed to fill and propel the shells, and the Krupp company manufactured weapons, battleships, barbed wire and the other metal products needed for the war. Lloyd George remarked that Germany was the best organised country in the world and her organisation had told.

Germany's substantial industrial capacity and technological capability gave the country a head start in the war. On the Eastern Front, German guns outnumbered Russian guns by almost two to one per division. In East Prussia, the Russian Tenth Army was starved of weapons and ammunition, most notably in the Second Battle of the Masurian Lakes which was fought in atrocious weather conditions in February 1915. The Russians needed some 45,000 shells per day to fend off the attacks by the German Eighth and Tenth armies; however, Russian factories could only supply 20,000 per day and the Germans eventually destroyed the bulk of the Russian Tenth Army. Around 56,000 Russian soldiers died, even more were taken prisoner and some 300 guns were lost.

On the Western Front, the French Army had only 300 heavy guns compared with 3,500 German medium and heavy guns when war broke out in 1914. The story was much the same in Britain: in May, when the Germans were turning out a quarter of a million of high-explosive shells daily, Britain was producing only 2,500, plus 13,000 shrapnel shells, as Lloyd George told the House of Commons in December 1915.

The British shell shortage led to the so-called 'shell scandal' that erupted on 14 May 1915. It was sparked by Sir John French (1852–1925), the commander-in-chief of the BEF, when he commented to Colonel Charles Repinton, *The Times* war correspondent, that the Battle of Neuve Chapelle had failed because of a lack of shells: 'The want of an unlimited supply of high explosives was a fatal bar to our success.' Whereas Germany had huge stocks of shells, the BEF commander-in-chief had to ration the number of shells fired by British heavy guns to eight per day and by field guns to ten per day 'for ordinary purposes'. Furthermore, the BEF had only ten heavy guns per division whereas the German armies had twenty per division in the first half of 1915.

The battle, which took place from 10–13 March 1915, was aimed at breaching the German line at the village of Neuve-Chapelle, before driving on and seizing the nearby Aubers Ridge, and finally threatening Lille some 15 miles away. The British unleashed a forty-minute bombardment and quickly secured the village, although German defences then halted the advance and the British had to abandon their attempt to capture Aubers Ridge. Two months later, on 9 May, the British resumed their attack on Aubers Ridge but once again failed. The artillery only had enough shells for a short forty-minute bombardment. Moreover, 90 per cent of the shells were shrapnel shells which were ineffective at piercing the heavily fortified German defences. The attack was called off the following day, with British casualties amounting to around 11,500.

The British shell scandal resulted in the fall of the Liberal government led by Prime Minister Herbert Asquith (1852–1928), who then formed a coalition government and appointed David Lloyd George as Minister of Munitions. Lloyd George galvanised the British munitions industry, setting up a system of national munitions factories and purchasing machinery needed for its factories from the United States. Hundreds of private factories not previously engaged in the manufacture of munitions also co-operated in the scheme and within a few months the shell shortage had been overcome. A similar transformation occurred in France where, by the end of 1915, the country was producing virtually all the armaments and ammunition it needed. By 1917 the British Ministry of

Munitions was operating some 200 factories, manufacturing not only explosives and shells, but also aircraft and other products needed for the war effort.

The factories that assembled shell components and packed the shells with explosives were known as 'filling factories', and one of them, National Filling Factory No. 6, at Chilwell in Nottinghamshire, filled its first shell in January 1916. By 1918, it had filled over 19 million shells, as well as some 25,000 sea mines and 2,500 aerial bombs. Tragically, however, in July 1918 a large part of the factory was destroyed when 8 tons of TNT exploded, killing 137 people.

A variety of explosives

A military or naval artillery shell is a projectile, an object that is projected from a weapon, but unlike other projectiles used in warfare, such as bullets, cannon balls, shot or even arrows, all of which are solid objects, shells have hollow bodies that carry other materials, most notably explosives.

The explosives carried by shells in the First World War included low explosives, usually gunpowder, and high explosives such as lyddite, and were employed in various roles. They were often used as fragmentation bursting charges to burst the shell when activated by a fuse, and were known as 'bursting' or 'explosive' shells. The force of the burst and the hot, sharp metal shell fragments and splinters were intended to kill or maim enemy troops and destroy enemy defences and positions. Low-explosive bursting shells fragmented when the explosive ignited, while high-explosive (HE) shells burst when the explosive detonated.

In shells that carried shrapnel bullets, poison gas or other payloads, a smaller amount of explosive was used to open the shell and expel the payload when the fuse activated. These expelling or bursting charges were not intended to fragment the shells but rather to deliver the payload and were known as 'carrier' shells.

Small amounts of explosives, also known as bridging explosives, exploders or boosters, were usually incorporated in high-explosive shells to boost the detonation of the high-explosive bursting charge. Some low-explosive bursting shells also contained a small amount of a low explosive with a finer grain than that of the bursting charge. The fine-grained explosive was placed between the fuse and the bursting charge to ensure that the charge ignited.

Shell bursting charges and exploders are secondary explosives: they are ignited, in the case of low explosives, or detonated, in the case of high explosives, by a primary explosive in the shell fuse. A primary explosive is also known as an 'initiating' explosive, exploding when subjected to shock, friction or heat. In the fuse of a high-explosive shell, for example, the primary explosive explodes when the shell hits its target. Many fuses also contain a larger amount of explosive in their magazines to boost the explosion, so a fused shell might therefore contain several explosives: a detonator explosive and booster charge in the fuse, and an exploder and a bursting charge in the shell. The series of explosions from detonation in the fuse to explosion of the shell's bursting charge is known as the 'explosive train' of the shell.

Explosives used to propel shells out of weapons are known as propellants. Cordite, a low explosive, was commonly used for this purpose by the British in the First World War. The propellants were packed into cartridges and an ignition charge of gunpowder was usually incorporated at one end of the cartridge to initiate the explosion of the cordite. Cartridge ignition charges were ignited by priming devices known as tubes or primers: a typical priming device contained a priming powder that was ignited by a shock-sensitive explosive in the device's percussion cap. The sequence of the explosions from percussion cap explosive to the priming powder in the device, to the ignition charge and propellant in the cartridge is known as the explosive train of the propelling charge.

The combination of a priming device, propellant cartridge, shell and fuse is known as a 'round' of artillery ammunition. Depending on the type of weapon and nature of the shell, a single artillery round might therefore require some eight explosive components: a percussion cap explosive, a priming powder in the primer, an ignition charge and propellant charge in the cartridge, a detonator explosive and booster explosive in the fuse, and a booster explosive and main bursting charge in the shell.

Artillery

The First World War has sometimes also been called an artillery war, the intense and often lengthy bombardments turning some of the battlefields into eerie moon-like landscapes with few identifiable features.[5] Remarkably, shellfire is thought to have accounted for 60–70 per cent of all the battle casualties.

A wide range of military and naval artillery weapons were employed during the war. The armies, for example, employed field artillery which moved with the troops and siege artillery which remained static. Field artillery included light field guns, trench mortars, howitzers and slow-moving heavy artillery drawn by teams of carthorses. Guns, howitzers, and mortars are generally distinguished from one another by the nature of their barrels and how they are loaded and operated.

Artillery guns are long-range weapons with long, horizontal or slightly elevated barrels. The range of the gun is increased by raising the barrels, the maximum angle being about 35°. Guns use a cartridge containing a single charge of propellant to fire shells at high muzzle velocities in flat trajectories at visible targets. Muzzle velocity is the velocity at which a projectile exits from the muzzle, the front or open end of a weapon.

Howitzers generally have shorter barrels than guns, that are elevated at angles of 45° or more and are lowered to increase range. They fire heavier shells at higher trajectories, but with lower muzzle velocities than guns, and were particularly useful, due to the steep angle of descent of the shell, against enemy trench systems. Also, unlike guns, the range can be extended by increasing the number of propellant charges.

Mortars are loaded with the barrel almost vertical and fire projectiles at high angles of elevation over a shorter range. The range is increased by lowering the barrel towards a minimum angle of 45° and by increasing the amount of propellant.

Shells and propellant cartridges are loaded into weapons in one of three ways depending on the type of weapon. If they are loaded into the muzzle, like the mortar, the weapon is known as 'muzzle-loading' (ML). If they are loaded into the breech, that is the back-end of the barrel where the gunners operate, the weapons are known as 'breech-loading' (BL). Early breech-loading guns recoiled significantly each time they were fired, and gunners had to push the gun back into position before they could re-load it and fire it again. Because the propellants were often loaded into the breech in bags, these guns were sometimes known as 'bag-charge guns'.

The development of quick-fire (QF) guns and QF howitzers overcame the recoil problem. These rapid-firing weapons, as they were also known, were designed with a hydraulic and spring mechanism that absorbed the recoil and enabled the barrels to return to their original positions automatically after firing. The breeches of QF weapons were also simpler to open and close than those of BL weapons and therefore could be reloaded and fired again much more quickly. Propellants and primers, contained in brass cartridge cases, were loaded into the breeches of QF weapons, either separately from the shells or fixed to them. QF artillery pieces could fire five or more times as many rounds per minute as their BL counterparts.

Howitzers used by the British Army in the First World War were described by their calibre, that is the diameter of the bore of the barrel and therefore the diameter of the shells they fired. For example, the BL 6in howitzer introduced by the British in 1915 fired 6in shells. Guns, on the other hand, were usually described by the weight of the shells they fired; for example, the QF 18-pounder gun, one of the most famous British artillery weapons of the war, fired nearly 100 million 18.5lb HE, gas, shrapnel and other types of shell during the war. That works out at an average of well over forty shells fired every minute for the whole duration of the war.

Common features of shells

In general, shells and their contents were made of materials, most notably metals and explosives, manufactured using chemical processes. Natural materials, such as wood and paper, were also used in shells, but they played relatively minor roles.

First World War shells varied immensely in type and function, including high-explosive bursting shells, shrapnel shells, gas shells, smoke shells, incendiary shells and star shells. The shells also varied not only in their chemical constitution, but also in their size, weight and method of fabrication. Yet despite the immense variety of shells fired in the war, there were a number of common features.

The empty shell

The empty shell, that is the shell of the shell so to speak, is sometimes referred to as the 'shell body' or the 'shell case'.

In some ways the shell case is similar in function to an animal shell. The latter, whether an eggshell or the shell of a crab, mussel or snail, is a hard rigid covering that supports and protects soft body tissue and organs inside the shell. The hardness of the covering derives from the chemical composition of the shell. The shell is made of a type of material known as a biocomposite, that is a composite of biological chemicals, carbohydrates and proteins for example, and naturally-occurring inorganic chemicals such as calcium carbonate. Likewise, the case of an artillery shell has to be strong enough to support and protect its hazardous and unstable contents as the shell is packed, when it is stored or being transported, when it is loaded into a weapon, and during its trajectory after it has been fired. The cases of fragmentation shells, however, must not be too strong as they need to fragment when the bursting charge explodes.

Like the animal shell, the strength of an artillery shell case depends on its chemical composition. In the nineteenth century, the cases of most artillery shells were made by pouring a molten metal, iron or steel, into a mould in a process known as casting. By the beginning of the First World War, cast steel shells were becoming obsolete, at least in the British Army.[6] Cast iron shells on the other hand were still manufactured and used to a limited extent. They were fired for practice from siege guns, siege howitzers, field howitzers and heavy field guns, but not from light field guns. Cast iron gas shells were also used by the British artillery from September 1915 onwards.

The cases of the vast majority of shells fired during the war were made of forged steel. They were forged by punching a cavity into a solid cast steel ingot and forcing it through a die to form the shape of the shell. The cases were then heated to a high temperature and allowed to cool slowly. This heat treatment process, known as annealing, improved the strength of the steel.

The strength of the cases, however, depended not only on how they were manufactured but also on the chemical composition of the metal used. Cast iron, for example, contains about 2 per cent carbon, making the shells hard and brittle. Most of the forged steel used to make British shells in the First World War contained between 0.35 per cent and 0.7 per cent carbon and small percentages of the chemical elements nickel, manganese and silicon.[7] Since the steel was stronger than cast iron, the cases of forged steel shells were manufactured with thinner walls, allowing more explosive to be packed into shells of the same size.

Bursting charges and exploders

Gunpowder was used in the common shells employed by the British artillery at the start of the Great War, with finely grained gunpowder, inserted between the fuse and the bursting charge, used to initiate the explosion of the bursting charge.

The earliest high-explosive shells fired by the artillery in the war employed lyddite as a fragmentation bursting charge. This dark yellow material is a form of picric acid, a chemical also known as trinitrophenol, which has been melted and then allowed to solidify. The detonation of lyddite was boosted by incorporating an exploder in the shell. The exploder was typically picric powder or TNT powder, also known as trotyl. Picric powder is a pale yellow crystalline material formed from saltpetre, also known as potassium nitrate, and ammonium picrate. The latter is prepared by the reaction of ammonia and picric acid, and, like picric acid, is also a yellow compound. Later in the war, amatol – a high-explosive mixture of TNT and ammonium nitrate – was used as the bursting charge for shells.

High-explosive shells were filled with as much bursting charge as was physically possible, while the amounts of bursting charges in other types of shell, gas or shrapnel shells for example, were much smaller. Fine-grain gunpowder was typically used as the bursting charge for shrapnel shells, whereas gas shells employed picric acid, TNT or amatol.

Fuses

Fuses, usually made of an alloy such as gunmetal or bronze, were attached to shells to control the detonation or ignition of the shells' bursting charges. Detonators inside many types of fuse contained a primary explosive that detonated on impact or after a set time; the detonation initiating the explosion of a booster explosive, typically gunpowder, contained in a compartment of the fuse known as the 'fuse magazine'. The explosion detonated high-explosive bursting charges or sparked the ignition of low-explosive bursting charges.

Three types of artillery fuse were used in the First World War: percussion fuses, also known as impact fuses; time fuses; and time and percussion fuses.

High-explosive shells were generally fitted with percussion fuses, with several designs used by the British Army. One of the simplest, known as the 'direct action' fuse, incorporated a copper disc supporting a firing pin – a steel needle – over a detonating explosive. When the shell on its trajectory struck a solid object, which could be the ground or a solid wall, the disc was crushed, forcing the needle to pierce and detonate the explosive. The resulting flash fired the booster charge in the fuse magazine, which in turn detonated the high-explosive bursting charge in the shell. Another type of percussion fuse, known as a 'graze fuse', worked on a different principle. When the shell was checked in flight, for example by grazing the ground or by passing through a thin wall, the deceleration threw the needle into the detonator or the detonator backward onto the needle, thus initiating the detonation.

Time fuses were designed to detonate after a pre-set flight time, enabling a carrier shell to eject its payload at some point along the shell's trajectory. Two types of time fuses were used during the war: igniferous and mechanical. Igniferous time fuses employed a magazine of explosive and a time ring that contained a train of slow-burning powder. Gunners could vary the time interval between firing the shell and explosion of the shell by rotating the fuse time ring and therefore the length of the train. When they fired a shell, the shock forced a pellet of detonator explosive onto a needle in the fuse detonator. The flash from the detonation ignited the slow-burning powder which, after the set time, fired the explosive in the fuse magazine. Mechanical time fuses, which tended to be more accurate than igniferous fuses, were produced by Krupp in Germany and used a clockwork mechanism: after a set time, the mechanism released a firing lever which fired the detonator.

Some fuses, known as time and percussion fuses, detonated either after a pre-set time interval or if the shell grazed an object. Direct action, time, and time and percussion fuses were generally screwed into the noses of shells, while graze fuses, on the other hand, were designed to screw either into the nose or the base of a shell, depending on the type of shell.

Fabrication of a fuse for a First World War shell typically required fifty or more components, including, depending on the type of fuse, not just the firing pins and time rings but also plugs, springs, spindles, bolts, collars and hammers. Some of these parts, particularly items like washers, were made from materials such as paper, cloth or leather. However, the vast majority of fuse components were manufactured with precision engineering from brass, steel, phosphor bronze and other alloys as well as from metallic elements such as aluminium, tin, copper and lead.

To work, they also needed detonating and boosting explosives. Mercury fulminate, a chemical compound that contains carbon, nitrogen and oxygen, as well as mercury, was widely used as a detonating explosive, either by itself or in a mixture with other materials. Other detonating compositions used in the war contained potassium chlorate, a compound of potassium, chlorine and oxygen. Some types of fuse employed 'mealed powder', a dust-like form of gunpowder that readily explodes when struck by a firing pin, to fire the booster explosive in the fuse magazine.

Fuse magazines typically contained loose, fine-grain gunpowder, with much larger particles than those of mealed powder, as the booster explosive. A form of gunpowder, known in the war as 'black powder', was used as the slow-burning powder in some time fuses and consisted of seventy-five parts of saltpetre, also known as potassium nitrate, fifteen parts of charcoal, and ten parts of sulphur. An explosive, referred to as 'composition exploding', was sometimes employed in place of fine-grain gunpowder as the booster explosive in fuse magazines in some types of high-explosive shell. The composition was a chemical compound, commonly referred to as 'tetryl', with several complicated chemical names: two of them are trinitrophenylmethylnitramine and *N*-methyl-*N*-2, 4, 6-tetranitroaniline.

Propellants

Propellants are simply materials that propel objects or other substances by creating thrust. When children propel peas from pea shooters, they use the breath exhaled from their lungs as a propellant, while blowpipes, also known as blowguns, work on the same principle. Indigenous peoples in South America and other regions of the world employ these weapons to hunt animals; their ammunition consisting of darts with poisonous tips. Similarly, cans of aerosol spray contain liquid propellants that release pressurised vapours when their valves are opened, propelling the liquid products, typically deodorants, out of the cans. Whereas blowpipe and aerosol-can propellants are gases and liquids respectively, the propellants used to fire bullets and shells from weapons are solids. Their propulsion relies on rapid combustion of the solids to generate gases under pressure within the confines of a firing chamber.

Combustion is a type of chemical process known as oxidation. When oil burns in a lamp or a log in a fireplace, oxygen from the air combines with the carbon and hydrogen in the oil or wood to form carbon dioxide, water vapour, a variety of gases and lots of energy. Oil and wood are fuels because they store energy, and the energy of a fuel is released by its reaction with an oxidant, which, in the case of combustion, is oxygen. However, when solid chemical propellants, such as gunpowder, burn in a sealed firing chamber, oxygen from the air is not available to support the combustion. The chemical propellant therefore has to supply its own oxygen.

Gunpowder, for example, comes in various forms but is typically a mixture of carbon and sulfur, both of which are fuels, and potassium nitrate which is the oxidant. When the powder burns, the oxygen in the nitrate oxidises the carbon and sulfur. The process not only produces lots of energy, but also results in the rapid formation of gases, most notably nitrogen, carbon monoxide and carbon dioxide. The gases rapidly generate pressures in the firing chamber that are high enough to propel a bullet or shell out of a barrel at high speed. The process also creates a variety of smoke-forming solid particles.

Cordite was the most widely used British artillery propellant in the First World War. The propellant varied in its formulation, but typically consisted of 65 per cent by weight of guncotton, 30 per cent nitroglycerine and 5 per cent petroleum jelly. The mixture was pressed into the form of cords.

Guncotton, also known as pyroxylin, nitrocellulose or cellulose nitrate, is a highly flammable yellowish-white compound consisting of filaments that look like raw cotton, and is made by the reaction of nitric acid and cotton. The material, a high explosive, combines a fuel and an oxidant in the same compound. The fuel is the cellulose part of the compound and the nitro part is the oxidant. The compound burns to form nitrogen, carbon monoxide, carbon dioxide and other gases. Guncotton formed the basis of '*Poudre B*' (Powder B) and subsequent '*poudre*' propellants developed by the French in the late nineteenth and early twentieth centuries.

Nitroglycerine is a pale yellow oily liquid. The compound is known by several other chemical names, such as glyceryl trinitrate. Like guncotton, it is a high explosive that combines a fuel and an oxidant in the same compound. Guncotton and nitroglycerine are both unstable and liable to decompose or explode under certain conditions. Nitroglycerine in its pure form, for example, detonates when subjected to physical shock. The petroleum jelly was therefore added to stabilise the cordite mixture and keep it free from moisture.

Ballistite, another propellant used during the war, consisted of 50 per cent nitrocellulose and 50 per cent nitroglycerine. A small amount of camphor or aniline, both of which are organic compounds, was sometimes added to the mixture to stabilise the two high explosives.

Even though cordite, the French *poudres* and ballistite contained high-explosive components, the propellants are classified as low explosives since they burn rather than detonate when ignited. They are also known as smokeless powders because their combustion produced greater amounts of gaseous products and far fewer smoke-producing solid particles than gunpowder. These propellants were loaded into weapons in various ways depending on the type of weapon. For trench mortars, cartridges consisting of one or more rings of the propellant charge were attached to the bases of the shells. The rounds were then dropped base first into the muzzle of the mortar. The cartridges also contained a small charge of gunpowder to ignite the propellant and a percussion primer that ignited the gunpowder.

For BL guns and BL howitzers, propellant charges contained in combustible fabric bags were loaded into the breeches of the weapons after the shells had been loaded. The bags were commonly made of silk or a tightly-woven, twilled woollen fabric known as 'shalloon'.

Bagged cordite cartridges contained bundles of cordite sticks tied up with threads. Some types of bagged cartridge consisted of doughnut-type rings of cordite tied around a central core of cordite. As cordite and other smokeless powders were difficult to ignite, an igniter pad which contained an ignition charge of fine-grain or sulfur-free gunpowder, or sometimes guncotton, was stitched into one end of each bag. The fabrics allowed the flash of flame from an ignition tube to penetrate the bags and ignite the cartridge.

A type of ignition tube, known as a vent sealing tube, was used to fire the bagged propellant. A typical vent sealing tube consisted of a brass body containing fine-grain gunpowder and gunpowder pellets. It was inserted into a vent that ran through the centre of the breech block. The fine-grain gunpowder was ignited, either electrically or by the action of a firing pin on a percussion cap, when the breech was closed. The explosion of the powder not only released a flash of burning pellets that travelled rapidly down the vent and hit the igniter pad of the bagged cartridge, it also caused the body of the tube to expand against the walls of the vent, thereby sealing it.

The sealed vent together with a pad in the breech mechanism prevented gases escaping through the breech after the propellant had been fired. The process is known as 'obturation', which means blocking or obstructing, and the pad was known as an 'obturator pad'. Breech obturation was necessary to ensure that the maximum possible pressure of the propellant gases was generated for propelling the shells out of the weapon. Any leakage of the gases reduced the pressure inside the weapon's firing chamber and therefore reduced the weapon's effectiveness.

Whereas BL weapons required bagged propellant cartridges, QF weapons used brass cartridge cases to contain the propellant, the cases acting as obturators. When the cartridge was fired, the case expanded and sealed the breech. The cartridges were either attached to the shells and loaded in one operation or loaded separately. After firing, the empty cartridge case could then be reloaded with propellant and used again several times.

A typical QF cordite cartridge contained one or more bundles of cordite sticks tied together and wrapped in silk-cloth bags. A percussion primer was screwed into the base of the brass cartridge case. The primer contained a magazine filled with fine-grain gunpowder and was fitted with a percussion cap containing either mercury fulminate or a flash powder mixture consisting mainly of potassium chlorate and antimony trisulfide: the potassium chlorate acting as the oxidant and the antimony trisulfide the fuel. When the breech was closed, the cap was struck and the resulting flash from the cap ignited the gunpowder which then fired the propellant cartridge.

Common shells

The original British 'common shell' was a type of bursting shell completely filled with gunpowder, usually a mixture known as 'P mixture', as a fragmentation bursting charge. The mixture consisted of cubes of gunpowder, known as 'pebble powder' and fine-grain gunpowder. Common shells had holes in their noses to take percussion fuses and several small bags of fine-grain gunpowder 'primers' were placed between the fuse and the bursting charge. The shells were made of cast iron, cast steel or forged steel and designed to rupture when the fuse activated on impact.

Common shells were phased out by the British Army for active service in the field at the beginning of the First World War in favour of high-explosive lyddite shells. However, cast iron shells continued to be used for practice firing of siege guns and howitzers.

The so-called 'common-pointed shells' were made of either cast or forged steel and filled with a larger charge of gunpowder than that used for the common shell. Common-pointed shells had solid noses, their bases were fitted with percussion fuses, and they were generally primed with several bags of fine-grain gunpowder. They were used by the British Royal Navy for firing at enemy ships, despite not being armour-piecing, and were also used to a limited extent by the British Army in the field, being fired from BL 9.45in siege howitzers.

High-explosive shells

The 'common lyddite shell' was the earliest type of high-explosive shell used in the First World War by the British Army, and was about four times more powerful than the gunpowder-filled common shell. The lyddite shell case was made of forged steel and molten picric acid was poured through the nose cavity almost filling the shell. The interior of the shell was varnished to prevent the reaction of the molten acid with the iron in the steel shell, a reaction which resulted in the formation of iron picrate. A bag of exploder was then inserted into the shell, either into a cylindrical cavity that ran through the centre of the lyddite or on top of the lyddite. An impact fuse was fitted into the nose of the shell.

The most effective lyddite shell explosions scattered shell fragments over a wide area. The effectiveness could be judged by the colour of the smoke released when the shell burst. Complete explosions were accompanied by black or grey smoke, or, when there was moisture in the atmosphere, an almost white smoke owing to the presence of steam. Yellow smoke contained unexploded particles of the yellow lyddite explosive and therefore indicated that the explosion was incomplete.

Amatol replaced lyddite as the main filling for British high-explosive shells in 1917, while some types of naval high-explosive shells employed TNT as the bursting charge.

Armour-piercing shells

Armour-piercing (AP) shells were made of cast or forged steel with thicker and therefore stronger noses and walls than other types of shells. Impact fuses, with bodies typically made of aluminium bronze, a hard and forgeable alloy of copper and aluminium, were fitted into the bases of the shells. In order to increase the strength of the shells, relatively small amounts of bursting charge were used compared with common-pointed and other types of shells.

Even smaller amounts of bursting charge were used in the stronger AP shell with cap, sometimes abbreviated to the 'APC shell'. The cap was made of mild steel which prevented the shell from shattering when it hit armour and enabled the shell to pierce armour with more force than the AP shell.

Bags of large-grain gunpowder were used as the bursting charges in early AP and APC shells, with lyddite subsequently used as a bursting charge for later APC shells. The interiors of these APC shells were lined with an aluminium container to take the high explosive.

Shrapnel shells

The shrapnel shell was invented in 1786 and named after its inventor Henry Shrapnel (1761–1842), an officer in the British Army's Royal Artillery. The shell case was a hollow cast iron sphere, fitted with a time fuse and filled with metal balls and a bursting charge of gunpowder. When the gunpowder ignited, the shell broke open and ejected the balls at speed. The shell was designed to explode in the air and shower enemy troops with the balls. The British Army adopted the shell for its artillery and used it in action in the early 1800s.

The cases of the shrapnel shells used in the First World War were made of forged or cast steel. The heads of most of the shells were fabricated separately and attached to the shell cases so that they could be readily blown off. They were filled with wooden blocks and fitted with time fuses or time and percussion fuses.

The shell cases were designed to carry as many balls as possible. The relatively small 13-pounder QF shrapnel shell, for example, carried 236 balls; the 60-pounder BL shrapnel shell, on the other hand, carried 990 balls. The balls were made either of cast iron or an alloy consisting of mainly lead and a small proportion of another chemical element – antimony. Antimony was used to harden the balls as those made of lead alone were considered to be too soft.

To maximise the effectiveness of the shells, the balls were made as small as possible and typically comprised almost 50 per cent of the weight of the shells. Molten resin was poured into the shells around the balls to stop them shifting around when the shells were in flight. The interiors of the shells were then lined with brown paper to prevent the resin sticking to the shell cases.

The fuse in the nose of the shrapnel shell was connected to a bursting charge of fine-grain gunpowder in a tin-plated iron cup in the base of the shell by a tube that passed down the centre of shell through the balls, the tube containing pellets of gunpowder. Many shrapnel shells also had a brass primer, filled with mealed powder, between the bottom of the fuse and the top of central tube. Detonation of the fuse fired the primer and the resulting flash passed down the central tube igniting the bursting charge. The explosion blew the head off the shell and expelled the shrapnel balls, also known as bullets, at a velocity greater than that of the shell. The term 'shrapnel' has also been commonly used to refer to the sharp, jagged metal fragments of shells that have burst apart.

Gas shells

During the course of the First World Was, gas shells were developed for various calibres of field artillery. Although the shells inevitably varied immensely in design and contents, they generally worked on the same principle. The shell cases were usually made of steel and filled with liquefied gases that readily vaporized to form harmful gases when the shells burst. The fillings included: tear gases, also known as lachrymatory chemicals; lung irritants; blistering agents known as vesicants; and other highly toxic chemicals.

Each shell was fitted with a fuse designed to set off a high-explosive bursting charge such as TNT or picric acid. When the shell exploded, it expelled the liquid into the atmosphere where it evaporated. Artillery shells filled with disabling or lethal chemicals were only used to a limited extent in the early part of the war. It was not until 1916 that chemical shells began to be used extensively by the British, French, Russian and German forces.[8]

One of the earliest types of chemical shell, the so-called 'T-shell' was developed by the Germans in 1914 for its heavy field howitzers.[8] The shell contained a mixture of two bromine-containing organic lachrymatory chemicals: xylyl bromide and benzyl bromide. The Germans called the mixture 'T-Stoff'. Like other chemical warfare agents employed in gas shells, T-Stoff could be readily liquefied at moderate pressures and ambient temperatures and was therefore easily compressed as a liquid into a shell. A lead receptacle for the T-Stoff was placed inside the steel bodies of T-shells, which stopped the T-Stoff corroding the steel case and therefore prevented premature release of the mixture. A high-explosive bursting charge and fuse were then located in the head of the shell above the receptacle.

On 31 January 1915, the German artillery fired some 18,000 T-shells at Russian troops on the Eastern Front during the Battle of Bolimów, near Warsaw. However, cold weather prevented the T-Stoff evaporating sufficiently and the shells therefore proved ineffective. Even so, the German Army continued to use the shells during warm weather in other battles on the Eastern and Western Fronts until 1917.[9]

The Germans also developed other types of field howitzer T-shells containing more volatile mixtures of lachrymatory organic chemicals that could be used in winter. They also developed trench mortar gas shells containing lachrymatory compounds known as 'B-Stoff' and 'Bn-Stoff', which were relatively simple bromine-containing organic chemicals and also known as bromoacetone and bromomethylethyl ketone respectively.

Similarly, the German C-shell, a trench mortar tear-gas shell, and K-shell, a howitzer tear-gas shell, were filled with the chlorine-containing chemical chloromethyl chloroformate, also known as 'C-Stoff' or 'K-Stoff'.

Chlorine and bromine are two members of a group of chemical elements known as halogens. In 1915 and 1916 the French introduced a variety of lachrymatory gases for their tear-gas shells, which were all organic chemicals containing a halogen. Because of the shell shortage, however, the British artillery did not start to use tear-gas shells to any great extent until 1917.[10] These shells were filled with ethyl iodoacetate and related halogen-containing tear-gas producing liquids, and TNT was used as a bursting charge. The shells were fired during the Battle of Arras in April 1917.

The first shells containing the highly toxic chemical carbonyl chloride, commonly known as phosgene, were fired by the French artillery in February 1916 during the Battle of Verdun.[11] Soon afterwards the Germans introduced shells marked with a green cross to indicate that the chemical contents of the shell were extremely harmful. 'Green Cross' shells were filled with phosgene or other toxic chlorine-containing organic chemicals.

The most effective lethal gas used in shells in the First World War was diphosgene, a chemical that has twice as many chlorine atoms in each molecule as its cousin phosgene. Like many organic chemicals, diphosgene has a variety of more lengthy chemical synonyms, one of them being trichloromethyl chloroformate.

Chloropicrin, also known as trichloronitromethane, was also used in German Green Cross shells, although it was initially used by the Russians on the Eastern Front in August 1916.[12] From early 1917 onwards, the British artillery also fired shells filled with phosgene or chloropicrin.

Blistering agents, also known as vesicants, were first employed in 1917. German 'Yellow Cross' shells, for example, were filled with mustard gas, a vesicant and lung irritant. Mustard gas, also known as dichloroethyl sulphide, is in fact an oily liquid and an organic chemical containing not only chlorine but also sulfur. The Germans first used mustard gas shells at Ypres in July 1917 against British troops, firing some 50,000 rounds on the night of 12 July.[13] The French started to produce mustard gas shells in March 1918 and began to fire them on the Western Front in June 1918. The British initially experienced problems in the production of mustard gas shells and were unable to fire any until the last two months of the war.

The United States also developed and manufactured gas shells, but none were fired during the war.[14] By the time the first consignment of American gas shells arrived in France, the war had ended. The United States artillery did, however, fire French-manufactured gas shells filled with phosgene and mustard gas.

Chemical shells contained not just liquids that vapourised to form harmful gases, but also contained harmful solids. The first-ever type of chemical-filled shell fired in the First World War was a type of shrapnel shell filled with balls embedded in a sternutatory powder, known as dianisidine chlorosulfonate, a type of irritant powder that causes sneezing, coughing, nasal irritation and tears. The German artillery fired several thousand rounds of these shells at British positions at Neuve Chapelle in the north of France on 27 October 1914. However, the powder proved ineffective and the use of the shells was abandoned.

From mid-1917 until the end of the war, the German artillery fired some 10 million shells filled with another type of sternutatory powder against American, British, French and Russian troops. The chemical fillings of these so-called 'Blue Cross' shells were typically arsenic-containing organic chemicals such as diphenylchlorarsine. Once again, the shells were not regarded as a success as they had little or no effect on the Allied troops.

Gas delivery systems

Artillery pieces of various types, principally trench mortars and howitzers, and a range of other delivery systems were employed by both the Allies and the Central Powers during the war to project chemical warfare agents at enemy troops.

In the early phase of gas warfare, the British Army started to employ the chemical mortar, a 4in version of the Stokes mortar, to project gas shells. The Stokes mortar, designed by Sir Wilfred Stokes in 1915, consisted of a 3in diameter steel barrel fixed to a base plate, supported at an angle on a bipod mount, and fired high-explosive shells.

However, 3in shells were not big enough to carry liquefied gas. The 4in Stokes chemical mortar could fire shells, or bombs as they were often called, containing poisonous gases such as phosgene or chloropicrin. The shell typically consisted of a cylindrical steel tube filled with the liquefied gas. An impact fuse was screwed into one end of the cylinder and a steel base disk fitted to the other. A central tube ran through the chemical filling between the fuse and base and contained a bursting charge, such as tetryl, and a detonator, often mercury fulminate. A cartridge containing a percussion primer, an ignition charge and propellant was attached to the base disk. The shell with the cartridge attached was then dropped through the muzzle of the mortar onto a firing pin at the base of the mortar barrel, detonating the priming powder in the cartridge which ignited the ignition charge and the propellant.

One of the most notable gas delivery devices was the Livens projector, a type of mortar invented in 1916 by English engineer Captain William Livens (1889–1964). He was an officer in the Royal Engineers, a division of the British Army also known as 'the Sappers'. The projector was used as a weapon by the British Army in 1917 and 1918 to fire large shells filled with phosgene, chlorine, chloropicrin, flammable oil, high explosives and other materials at German positions. The weapon consisted of a steel base plate and a 3–4ft long steel barrel with an 8in internal diameter, which was buried in the ground at an angle of 45°. Livens projector chemical shells were drums filled with 30lb of the liquefied poisonous gas. Each shell had a central tube fitted with a fuse, a primer and sufficient TNT bursting charge to split open the shell and discharge the toxic contents. The shells were fitted with handles that allowed the troops to lower them into the barrels onto bags of propellant charge, which was ignited electrically. The range of the projector was increased by increasing the number of charge bags.

Batteries of sometimes thousands of Livens projectors were able to propel tons of poisonous chemicals rapidly at a target at ranges of up to a mile, releasing immense clouds of toxic gases over the German frontlines. A special brigade of British gas troops made their first full-scale gas attack with these projectors in April 1917 at the Battle of Arras. The brigade launched its largest attack in March the following year when it fired some 3,730 Livens gas shells into the French city of Lens and its outskirts.

Shells were not the only means of projecting dangerous chemicals against the enemy, however. Military aircraft, for example, used bombs, sprinklers or sprayers to disperse the chemicals. Gas cylinders were also widely used on the ground, especially in the early part of the war. On opening a valve on top of the cylinder, gas under pressure was discharged through a pipe and nozzle towards enemy line.

In one of the most famous episodes of chemical warfare, the Germans released 168 tons of chlorine from 5,730 gas cylinders at Langemarck, near Ypres, on 22 April 1915. The cloud of gas was carried by the wind over trenches occupied by French troops, resulting in more than 3,000 casualties, including at least 800 fatalities. The success of this attack caused the Germans to subsequently release chlorine from cylinders in three other attacks in April and May 1915.

The first British gas attack also used cylinders of chlorine: between 5.40 a.m. and 6.30 a.m. on 25 September 1915, the British artillery released 140 tons of chlorine from over 5,000 cylinders at the Battle of Loos. At the same time they used mortars to fire some 10,000 smoke bombs filled with phosphorus, another chemical element.

The British broke through the German trenches and captured the town of Loos, but not without experiencing several problems with the gas. First, a variety of chlorine cylinders had been supplied to the British frontline, but not always with the correctly-fitting spanners to open the cylinder valves. Second, the joints of the lead pipes used to dispense the chlorine from the cylinders into the wind above the trench parapets leaked, allowing the gas to escape into the British trenches. The release of the gas through the open valves from the cylinders into the pipes caused the valves to freeze and so the troops could not close them. Next, the wind was not always strong enough to carry the gas over the German trenches, so some of it hung around in 'no-man's-land' and blew back over British trenches, gassing some of the soldiers with their own gas. Finally, German shells hit some of the full chlorine cylinders, releasing even more gas into the British trenches. The British carried out a second chlorine gas attack on 26 September and another on 13 October, the final day of the offensive.

Smoke shells

Smoke is a visible collection of enormous numbers of exceedingly small solid particles or liquid droplets dispersed in air. It is typically a product of the combustion of natural materials such as wood, tobacco, coal and crude oil; however, it is also produced from a range of chemicals.

In the First World War, various types of smokes were generated artificially to provide cover for naval and military manoeuvres and operations. These obscurant smokes, as they are called, were used in two ways: discharged over and around enemy troops in order to unsight them, known as smoke blankets; and clouds of smoke generated close to and in front of friendly troops or ships to mask their activities, known as smoke screens.

On 31 May 1916, during the Battle of Jutland, the German High Seas Fleet, under the command of Admiral Reinhard Scheer (1863–1928), laid a smoke screen to provide cover for the fleet while it executed a turning movement to escape from the pursuing British Grand Fleet under the command of Admiral Sir John Jellicoe (1859–1935). The screen was created using two types of smoke: plumes of black smoke from the ships' funnels were generated by restricting the amount of air available to burn the crude oil fuel that was used to heat the boilers; white smoke was produced from two liquid sulfur-containing chemicals on board the ship and from floating containers. One of the chemicals, sulfur trioxide, reacted with the water in the humid air above the sea to form clouds of dense fog-like suspensions of minute droplets of sulfuric acid, and its cousin sulfurous acid. The other, chlorosulfonic acid, reacted with the moisture in air to produce fumes of hydrochloric acid and sulphuric acid. The Germans also used sulfur trioxide and chlorosulfonic acid to fill their military smoke shells, although the Allies principally used white phosphorus as their smoke generating material.

White phosphorus is one of two common forms of the chemical element phosphorus, which is a soft, waxy, flammable white solid. Although it does not dissolve in water, the element ignites spontaneously in air, combining with the oxygen to form phosphorus pentoxide, a white solid which in turn combines with the moisture in air to form phosphoric acid.

The other common form of the element, red phosphorus, is a red powdery solid and is also insoluble in water but reacts readily with oxygen under normal conditions. Unlike white phosphorus, red phosphorus was rarely used by itself in smoke shells, although mixtures of the two were sometimes used.

White phosphorus is manufactured from phosphate rocks such as apatite which contains calcium phosphate and related phosphorus-containing chemical compounds. The white phosphorus can then be converted into red phosphorus. Phosphate rocks are found in the Middle East, North Africa, the United States and parts of Europe such as France and Belgium. They were therefore available in abundance to the Allies during the First World War; in Britain alone, annual phosphorus production increased from approximately 1,000 tons in 1914 to 2,500 tons in 1918.[15]

However, phosphate rocks were not available to such an extent to Germany and Austria. The Central Powers could therefore only manufacture limited supplies of white phosphorus and, as a consequence, had to rely on the use of sulfur-containing compounds for production of their obscurant smokes.

The obscuring power of smokes generated from white phosphorus is some 50 per cent better than that of smokes generated from sulfur trioxide. The availability of white phosphorus for smoke shells and the superiority of the element as a smoke-producing material compared with sulfur-containing compounds gave the Allies a distinct advantage over the Central Powers when it came to the tactical use of smokes.

In the final stages of the war, the Americans produced smoke shells, some filled with white phosphorus and others filled with titanium tetrachloride as the smoke-producing chemical. Titanium tetrachloride, a non-flammable but highly corrosive liquid, is an irritant to eyes and the respiratory tract. The compound absorbs moisture from the air to form dense white fumes of hydrochloric acid and titanium compounds such as titanium oxychloride.

Smoke shells were generally designed in similar fashion to gas shells. They were fitted with a fuse, had a high-explosive bursting charge, and contained as much smoke-generating material as technically feasible. Corrosion-resistant lead receptacles were sometimes placed inside the shells to contain corrosive chemicals such as sulfur trioxide.

Smoke shells were not just used to generate obscurant smokes on the battlefield but also as practice shells. The British War Office *Treatise on Ammunition* published in 1915 described, for example, a practice shell for the QF 4.5in howitzer.[16] This howitzer fired 35lb high-explosive, shrapnel, smoke, gas and other types of shell. The practice shell contained a gunpowder bursting charge and a smoke-producing mixture of lead sulfate, tin oxide, aluminium powder and naphthalene, which vapourised to form a dense white cloud of smoke.

Incendiary shells

The use of incendiaries in warfare dates back to several centuries BC when armies used resin-and-straw fire balls and burning oil to attack or defend fortified cities. In the Great War a variety of incendiary munitions were employed, including armour-piercing shells, grenades and aircraft bombs containing an incendiary material such as white phosphorus.

Another incendiary device, the notorious flamethrower, was developed in Germany before the start of the war. It used pressurised gas to project streams of burning fuel against enemy troops on the ground. The fuel was a mixture of hydrocarbons obtained by distilling petroleum or coal tar. Flamethrowers were first used in the war by the German Army in 1915, most notably on 30 July against British trenches at Hooge, a small village in the Ypres Salient, Belgium. They were subsequently used by the British and the French armies, although with limited success.

Incendiary artillery shells were developed by both the French and Germans before the war and by all sides during the war. In his account of an attack on the Western Front near Ypres in late July 1917, Jünger refers to incendiary shells throwing up 'heavy milk-white clouds, out of which fiery streaks dribbled to the ground.'[17] He watched a 'piece of that phosphoric mass' drop to a stone at his feet where it burned for several minutes.

Although incendiary shells were often directed against enemy troops, they were more commonly used to destroy buildings, equipment and protective cover. For example, incendiary shells were initially employed as anti-aircraft shells to set fire to observation balloons, which were filled with a highly flammable gas, hydrogen. Incendiary trench mortar shells and Livens projectiles were employed not only to demoralise the enemy during gas attacks, but also to clean out woods and destroy shrubbery, grass or camouflage that provided cover for enemy movements.

Two types of incendiary material were used as fillings for incendiary shells: spontaneously combustible materials and materials that had to be ignited before they burst into flame. White phosphorus was the archetypal spontaneously combustible material, used by both the Germans and the French in shells to deliver not only obscuring smoke but also to scatter burning particles of the element. An early type of German incendiary shell consisted of a cast iron case containing cylinders of celluloid and camphor, both highly flammable organic materials, surrounded by white phosphorus. Paraffin wax, another flammable material, was used to keep the cylinders in place inside the shell and black powder was used as a bursting charge.

German and British shells also used thermite mixtures as incendiary fillings. Thermite is a mixture of a metal powder and a metal oxide, usually finely powdered aluminium and iron oxide. When ignited, these two components react violently to form iron and aluminium oxide, the chemical reaction generating sufficient energy to melt the iron. It is used for welding steel, for example, and for joining railway tracks.

One thermite-containing shell, the British 18-pounder incendiary shell, employed a mixture of potassium perchlorate and magnesium powder to ignite the thermite. The mixture, known as ophorite, was contained in a central tube in the shell. The ignition mixture was readily ignited by the flash from a fuse containing fine-grain gunpowder. The thermite ignition products, molten iron and hot aluminium oxide were expelled through the nose of the shell by a bursting charge of ophorite in the base of the shell. The shell functioned in a similar way to a shrapnel shell, the incandescent products were scattered over the target. They were sufficiently hot to penetrate metal and prolong the burning of combustible materials such as wood, which proved to be particularly effective against aircraft.

The largest incendiary shell used during the war was the German 175mm (6.9in) shell, which contained two incendiary materials: thermite and the metallic element sodium. The sodium, which was ignited by the thermite reaction, burnt in air to form sodium oxides. Sodium oxides react with water to produce highly caustic alkaline solutions of sodium hydroxide that can burn through clothes, skin and flesh. Sodium itself reacts vigorously with water to form not only sodium hydroxide solution but also highly flammable hydrogen gas. The reaction generates sufficient energy to ignite the gas.

Tracer shells

Tracer shells enabled gunners to trace the trajectories of the shells and therefore adjust their aim as necessary. Tracer-incendiary shells, for example, were often fitted with time fuses set to ignite the incendiary materials well before the shells reached their targets, typically balloons. The incandescent products were then ejected through the noses of the shells towards the targets. At the same time, the smoke from the combustion of the incendiary materials left visible traces for the gunners.

Some types of tracer shell contained a tracer device fitted into the base of the shell. The devices contained pyrotechnic mixtures similar to those used in fireworks, which produce intense light without exploding. The tracer mixture in the device was ignited by the burning propellant when the shell was fired.

The mixtures used in First World War tracer shells typically contained magnesium powder and a metal salt such as barium nitrate or strontium nitrate. Magnesium is easily ignited and burns with a brilliant white flame, while barium and strontium salts burn with pale green and crimson flames respectively. The devices emitted highly visible flames and smoke from the shells during their trajectories.

Star shells

Star shells, also known as illumination shells, were fired to illuminate the battlefield at night and therefore enable soldiers to see enemy troops and positions. They were also used for signalling purposes.

To some extent, star shells, which commonly contained ten or twelve stars, resembled shrapnel shells in design and function. A time fuse fitted into the nose of the shell was connected to a fine-grain gunpowder bursting charge. A typical star consisted of a rolled brown paper cylinder containing a pyrotechnic mixture of barium nitrate, potassium nitrate and magnesium powder. The cylinder also contained mealed powder and quickmatch threaded through the ends of cylinder for priming the pyrotechnic mixture. Quickmatch is a string-like fuse made by boiling a cotton yarn in a solution of mealed gunpowder and gum Arabic.

Star shells were designed to burst high in the air. As they burst, they ignited and scattered the stars which then burnt brightly and drifted slowly to ground. A variety of star compositions were also used for signal rockets and Very flares. Signal rockets typically contained twenty or thirty stars and functioned in the same way that firework rockets are fired nowadays. They propelled themselves through the air using gunpowder as a propellant, with quickmatch used as a primer.

Very flares, also known as Very lights, are named after their inventor, the American naval officer Edward W. Very. There are numerous references to the flares in First World War accounts; for example, in her book *The Roses of No Man's Land*, author and First World War historian Lyn MacDonald refers to Very lights blazing and sparkling in the sky above Hooge in 1916.[18] The lights were produced by firing cartridges from pistols known as Very pistols: each cartridge contained a single star, a percussion cap and a propellant charge of gunpowder. The different colours of the rocket and Very flares were obtained by varying the star compositions. When ignited, compositions containing magnesium powder, copper carbonate, barium chlorate and a strontium carbonate produced white, blue, green and red flares respectively.

3

Mills Bombs and other Grenades

Wonderful grenades

'Hold the grenade in the right hand ... Pull out the safety pin with the left hand' and throw the grenade with an overarm action 'like a bowler when playing cricket', notes Major Graham M. Ainslie in his handbook on hand grenades published in 1917.[1] Ainslie, a serving officer in the British Army in the First World War, observes that these instructions were 'the result of practical experience' in the war.

On 4 July 1916, during the First Battle of the Somme, British war poet and soldier Siegfried Sassoon (1886–1967) used hand grenades known as 'Mills bombs' to storm an enemy trench at Mametz Wood. Whether or not he used an overarm action is unclear. What is clear, however, is that Sassoon's bombing technique combined the practical method, subsequently recorded by Ainslie in his handbook, with a spur-of-the-moment improvisation of his own. In his memoirs, Sassoon describes running towards the enemy trench with a bag of Mills bombs slung over his shoulder and a Mills bomb in each hand.[2] As he ran, he pulled out the safety pin from a bomb in his right hand and used his teeth to extract the safety pin from a bomb in his left hand, clearing the trench of German troops with this unique two-handed approach. Sassoon, who was renowned for his daring, if not recklessness, during raids on enemy trenches, was awarded the Military Cross later that month for his gallantry on one of these raids.

There is little doubt that British soldiers found Mills bombs to be highly effective close-quarters weapons for storming German trenches. In *Old Soldiers Never Die*, Frank Richards (1883–1961), a private in the British Army, remarks that Mills bombs were 'wonderful for throwing into shell holes, trenches and dugouts' and other types of grenades were 'very crude' in comparison.[3]

Mills the inventor

Sir William Mills was a British foundry owner, an engineer and a keen inventor. He introduced the hand grenade that was named after him in February 1915, and the first orders for the Mills bomb, which the British Army adopted as its standard hand grenade, were placed in April and May that year. Between then and the end of the First World War, around 75 million Mills bombs of various designs were supplied for the use of the Allies.[4]

Mills, the son of a shipbuilder, was born in Sunderland, England, on 24 April 1856. In his twenties, he spent seven years at sea as a marine engineer repairing telegraph cables. During this time, he became increasingly interested in metallurgy and innovative engineering designs, and at the age of 29 he set up the first aluminium foundry in Britain at Monkwearmouth in Sunderland.

In 1892 and 1894 he patented an invention in Britain and the United States respectively for improving the gears used for raising and lowering ships' lifeboats, the invention was adopted by both merchant and naval vessels. An enthusiastic golfer, Mills became well known for the design and manufacture of some of Britain's first golf clubs with aluminium heads. Mills also established works in Birmingham and Sunderland to manufacture aluminium castings for motor cars and aircraft.

The grenade designed by Mills was not entirely unique, however. In 1912, Léon Roland, an officer in the Belgian Army, invented and patented a similar sort of grenade. Like the Mills bomb, the Roland grenade incorporated various features that prevented accidental detonation. Roland was taken prisoner by the Germans while on active service in November 1914 and the grenade named after him was never manufactured or used in warfare.

In January 1915, Albert Dewandre, one of Roland's engineering colleagues, and Mills got together to examine the Roland grenade. Dewandre and Mills both recognised that, although the grenade had a number of design defects, it had potential for trench warfare. After initially working with Dewandre on improvements, Mills collaborated closely on the design with the Royal Laboratory at Woolwich and William Morgan, an automobile engineer who was a professor at Bristol University.[5] By the end of 1915, he had filed nine patents, including two with Morgan, on improvements to the grenade.[6]

In the same year, Mills opened a munitions factory in Birmingham to manufacture the grenades. He claimed, however, to have lost money on the venture, although the British government are known to have paid him £27,750 for his invention, equivalent to approximately £2 million today.

Mills married Eliza, the daughter of a Manchester cotton spinner, on 10 November 1891 but they had no children. In 1922, he was honoured with a knighthood for his services. He died in Weston-super-Mare on 7 January 1932.

The Mills bomb

The pineapple-shaped Mills bomb used in the First World War had a body with grooves and rectangular segments that allowed it to be easily gripped. The body was made of cast iron which, unlike steel, readily fragmented when the bursting charge exploded. A filling hole in the body allowed the grenade to be charged with ammonal – a high-explosive mixture containing ammonium nitrate, aluminium powder and a small amount of carbon. After charging the grenade with the explosive, the hole was closed with a brass screw.

To arm the grenade the bomber unscrewed a plug at its base and inserted an igniter, which contained a percussion cap and detonator, made from nitroglycerine and mercury fulminate respectively, connected by a safety fuse. The base plug was then screwed back tightly into the grenade using a special key.

The percussion cap was located at the bottom of an aluminium or brass tube in the centre of the grenade. The cap was separated from a striker at the top of the centrepiece, as it was called, by an empty striking chamber. The striker, inside a type of spring known as a compression coil spring, was retained in position by the tip of a hand lever that was secured on the outside of the grenade by a safety pin. When secured, the bottom of the striker compressed the spring against the top of the grenade and therefore stored energy in the spring.

Before throwing the grenade, the thrower removed the safety pin with one hand and secured the lever against the grenade in the other. On tossing the grenade, the lever detached itself from the striker and allowed the spring to extend and release its stored energy. The lever flew away from the grenade while the released energy from the spring forced the striker down through the empty striking chamber and onto the percussion cap. The resulting flash from the cap ignited a slow match safety fuse, which was typically prepared by boiling cord made of hemp in a solution of potassium nitrate, also known as saltpetre. The cord was then dried. The type of slow match used for the Mills bomb igniter was designed to burn at 2ft per minute and so, as the slow match in the igniter was just 1.75in long, the fuse took about four-and-a-half seconds to burn. The automatic ignition of the safety fuse by a percussion cap was a significant advancement, as earlier time-fused grenades relied on the use of a match or lit cigarette to ignite the fuse.

The Mills bomb weighed around 1.5lb, and a reasonably strong and well-trained bomber could pitch the grenade up to 100ft. When the bomb detonated, it burst and scattered lethal cast iron fragments in all directions. The delay provided by the safety fuse allowed the bomber sufficient time to duck away and avoid being hit by any fragments.

Following its introduction in 1915, the Mills bomb underwent a series of modifications. For example, a vent channel was incorporated into the striker so that, when secured by the safety pin, the striker prevented air and moisture entering the grenade. When the striker hit the percussion cap, the vent opened up to allow gases from the combustion of the slow match fuse to escape into the empty chamber above the striker and then out of the grenade. The improvement reduced the number of premature detonations.

A version introduced in 1917 for use in hot climates was dipped in shellac after it had been filled with the bursting charge. Shellac is a natural moisture-resistant resin consisting of a natural polyester, a small amount of wax and a dye. The resin is secreted by parasitic insects that live in trees in Southeast Asia. Shellac coatings kept the Mills bombs dry and therefore prevented moisture penetrating the grenades and rendering them ineffective.

Other types of grenade

Before the introduction of the Mills bomb, grenades had required skilful handling as they could be quite dangerous in inexperienced hands. Some of the earliest versions tended to be quite crude. For example, iron spheres roughly 3in (7.6cm) in diameter filled with gunpowder and fitted with slow match fuses were used in the bloodless English revolution of 1688–1689 known as the Glorious Revolution.

Earlier in the seventeenth century, King Louis XIV of France established companies of elite soldiers, known as grenadiers, who were trained to throw grenades in their assaults on enemy troops. By 1672, every French infantry regiment had a company of grenadiers. In the following century, grenades were abandoned by the infantry in favour of new types of muskets with increased range, although companies of elite assault troops continued to be known as grenadiers.

Grenades were revived with the advent of trench warfare. They were employed, for example, by entrenched Russian and Japanese troops at the Siege of Port Arthur during the Russo-Japanese War of 1904–05. The Chinese deep-water port was heavily defended by the Russians but eventually fell to the Japanese.

It is noteworthy that a single hand grenade thrown in the Bosnian city of Sarajevo on the morning of 28 June 1914 precipitated the First World War. Nedeljko Čabrinović, a young Bosnian-Serb, tossed a percussion grenade at a car carrying Archduke Franz Ferdinand of Austria, the heir apparent to the Austro-Hungarian throne, and his wife Sophie on a visit to the city. The grenade bounced off the back of their car and exploded under the next car in the motorcade wounding a number of people. Later that morning, the Archduke and his wife travelled by car to visit the wounded in hospital. Before they could get there, however, they were shot dead by another young Bosnian-Serb, Gavrilo Princip.

Grenades soon proved to be useful weapons in the ensuing war. As Sassoon and Richards both observed in their accounts, hand grenades were particularly effective for raids, skirmishes and close combat on the frontline where no-man's-land often separated opposing trenches by just 300–400yds.

Numerous types of grenades were used during the First World War. Some, like the Mills bomb, had time fuses that delayed the explosion of the grenades, while others had percussion fuses which meant that they detonated on impact. Grenades were generally classified as hand grenades, stick grenades or rifle grenades: hand grenades were gripped in the palm of the hand and tossed like a ball; stick grenades had handles for throwing; and rifle grenades were fired from rifles.

The German Army adopted grenades as standard issue prior to the war. The BEF, on the other hand, had to rely on limited supplies and in the first few months of the war was supplied with no more than a few hundred crude percussion-fused stick grenades per month. This type of grenade had a cylindrical brass body attached to a cane handle. The brass cylinder was filled with lyddite, a high-explosive bursting charge, and a hollow pellet of compressed picric acid was placed in a cavity in the centre of the lyddite. The stick grenade, which weighed about 2lb, carried silk braid tail streamers to stabilise its flight and help it fly head first when thrown. A cast iron ring around the top of the brass cylinder made the grenade top heavy and, together with the streamers, ensured that the nose rather than the handle hit the ground first.

Before use, the thrower removed a cap on the top of the grenade, inserted an impact detonator and replaced the cap. The detonator device was a brass holder containing mercury fulminate, a primary explosive, and a needle secured by a phosphor bronze safety pin. Immediately before throwing the grenade, the bomber pulled out the safety pin. When it landed, nose first, the impact drove the needle into the primary explosive and detonated it, which detonated the picric acid pellet and then the bursting charge.

After removing the safety pins, these grenades were sometimes detonated accidentally as they were thrown. A soldier could, for example, if not careful, inadvertently knock the grenade against a trench wall or some other hard surface. Because of the lack of availability of these stick grenades in the early days of the war and their dangerous nature, British frontline troops often resorted to the use of improvised grenades with the safer time fuses.

Improvised 'jam tin' bombs of various designs, for example, were widely used in 1915. These grenades essentially consisted of empty jam tins filled with a high explosive such as ammonal and metal scraps, stones, ball bearings or other hard objects. Safety fuses, known as Bickford fuses, were inserted into the jam tin tops. A Bickford fuse consisted of a waterproof tube containing a train of fine-grain gunpowder enclosed in jute yarn. The fuse could be ignited by a lit cigarette or match and burnt at the rate of about 2ft per minute. The length of delay between lighting the fuse and the grenade exploding, typically a second or two, could be adjusted by cutting an appropriate length of fuse.

The most famous German First World War grenade was the *Stielhandgranate*, or 'potato masher' as British troops called it. This stick grenade, introduced in

1915, consisted of a hollow wooden handle attached to a thin steel cylinder containing a bursting charge and a detonator with a friction igniter and time fuse. A pull cord that ran through the handle was connected to a steel rod with a rough surface inside the igniter. When the thrower pulled the cord, the rod was pulled through a primer explosive in the igniter. The friction of the roughened surface ignited the primer in similar fashion to lighting a match by dragging the striking surface of a matchbox over the match head. The flash from the primer ignited the time fuse, which burnt for a few seconds before detonating the bursting charge.

Rifle grenades were propelled from a rifle using a blank cartridge, a cartridge containing a propellant, typically cordite, but no bullet. The Hales rifle percussion grenade was one such grenade used by the British Army in the war. The grenade was named after its inventor Marten Hale, an explosives expert and director of the Cotton Powder Company which produced guncotton from cotton waste, sulfuric acid and nitric acid at its factory near Faversham, England. Hale also invented a percussion stick grenade.

His rifle grenade consisted of a steel cylinder attached to a steel rod, the cylinder containing the bursting charge, detonator and striker. Before firing the grenade, the rod was oiled and inserted into the barrel of the rifle. A safety pin that locked the striker in place was removed from the grenade before squeezing the trigger of the rifle. On firing, the propellant gases from the blank cartridge propelled the grenade and rod, which acted as a tail, through the air. On impact with a hard surface, the striker struck the detonator which then detonated the bursting charge.

Ainslie's handbook outlines the main features of and firing instructions for more than twenty British, French and German hand, stick, and rifle grenades. Whereas many of these grenades employed ammonal as the bursting charge, other types of explosive were also used. For instance, the Hales rifle and stick grenades contained tonite, a high-explosive mixture of guncotton, also known as nitrocellulose, and barium nitrate as the bursting charge.

Chemical grenades

The vast majority of the grenades used in the First World War were high explosive fragmentation grenades designed to kill or maim enemy troops by bursting into lethal fragments. Some, however, were designed to deliver chemical agents. The French, for example, used chemical grenades against the Germans in August 1914 – the first month of the war. The grenades were filled with xylyl bromide, a toxic organic compound that exists as a liquid under normal conditions. The chemical is a lachrymatory agent, meaning that it vapourises to form an eye-irritant vapour, more commonly known as tear gas. In November

1914, the French used rifle and hand grenades to deliver ethyl bromoacetate and chloroacetone, both of which are also lachrymatory agents.[7]

Chemical grenades typically consisted of a steel body filled with the chemical agent, a detonator, a Bickford safety fuse, and a firing mechanism with a lever and safety pin.[8] Once the pin was removed, a striker pin struck a priming powder composition at the tip of the fuse. The flash from the primer ignited the fuze which burnt for five seconds and then exploded the detonator. The grenade burst, scattering fine droplets of the chemical agent, if it were a liquid, or fine particles in the case of a powdered chemical agent.

The fillings included not only lachrymatory agents like those used by the French early in the war, but also lethal chemicals such as phosgene, smoke-producing chemicals such as tin tetrachloride, and incendiary materials such as thermite. Phosgene, known also by its chemical name carbonyl chloride, is the most notorious of lethal chemical agents; tin tetrachloride, or stannic chloride, is a liquid that evaporates to form an obscurant smoke when it comes into contact with air; and thermite is a mixture of aluminium powder and iron oxide that reacts and burns at a high temperature when ignited.

The Allied and Central Powers both used chemical grenades in the war. The British infantry, for example, used various types of chemical grenades during the 1917 Battle of Ypres, also known as the Third Ypres, that was fought from 31 July to 10 November. They included phosgene grenades, stannic chloride grenades and white phosphorus grenades.[9] White phosphorus is both a smoke-producing material and an incendiary material when it comes into contact with air.

4

The Highs and Lows of Explosives

Enormous quantities of explosives

> I'd kill all chemists and explosive experts!

This outburst, uttered by an un-named British officer during the First World War, was not so much a threat but a cry of anger and despair. It came while the officer was commenting on the development of weapons in the war, lamenting that modern warfare was 'a dirty scientific business'.[1]

This scientific business, as he put it, or at least its application by the military, led to killing and maiming on an industrial scale. Casualty statistics for the war are difficult to calculate because accurate data were not always collected. Even so, there is general agreement that upwards of 6 million men in the forces of the Allies and Central Powers were killed in battle, millions of others died in captivity or from disease and an estimated 20 million or so men were wounded. These estimates do not include the almost 7 million civilians who met their deaths as a result of the war.

There is no doubt that much of this industrial-scale killing and maiming would not have occurred without the development of efficient chemical processes before and during the war for the industrial-scale manufacture of chemicals such as nitrocellulose, nitroglycerine, picric acid and trinitrotoluene. These chemicals were needed to fill and propel the shells fired by the opposing armies during the war.

The British Army and Royal Navy together fired almost 250 million shells during the war. That works out at an average of about 100 shells per minute throughout the war. According to military historian Philip Haythornthwaite, an average of one gun every 30yd 'was regarded as normal' on the Western

Front which stretched for about 450 miles from the North Sea to the Swiss border.[2] During an offensive, the average increased to about one gun every 10yd.

The amount of explosive in each shell varied immensely. The 60-pounder BL field gun, for example, fired 60lb high-explosive and shrapnel shells as well as smoke and gas shells. Each high-explosive shell was filled with a high-explosive bursting charge, such as lyddite, and a booster explosive, such as picric powder, with a combined weight of 4lb.[3] The 60lb BL shrapnel shell, on the other hand, had a bursting charge of 4.5oz of fine-grain gunpowder and also contained 990 metal balls, weighing a total of around 36lb.[4]

The 60-pounder gun entered service in the British Army in 1905. Over the next thirteen years 1,756 BL 60-pounders were produced, of which 571 remained in service in 1918, firing a total of over 10 million 60lb rounds on the Western Front.[5] Each round required explosives not only for the bursting charge and exploder in the shell, but also for the propellant charge and fuse. That amounts to millions of pounds of explosive just for the BL 60-pounders over the course of the war.

Other guns used huge amounts of explosives during the war, such as the quick-fire 18-pounder field guns which alone fired almost 100 million rounds on the Western Front.[6] Add to that the explosives needed to propel the bullets from pistols, rifles and machine guns and the explosives needed for mines, grenades, bombs, torpedoes and demolition, massive amounts of a variety of explosives were manufactured to meet these needs during the war.

Cordite, which was used as a smokeless propellant by the British Army and British Royal Navy, is a prime example of the scale of explosives production. The largest cordite factory in the British Empire was located at Gretna, a town in Scotland that lies on the border with England. In 1917, production of cordite at His Majesty's Factory, Gretna, reached a peak of an astonishing 800 tons per week.

In North Wales, a plant at Queensferry, near Chester, was able to produce 500 tons of TNT and 250 tons of nitrocellulose per week.[7] Other plants focused on filling munitions with these explosives; for example, the National Filling Factory No. 6 at Chilwell in Nottinghamshire, England, filled more than 19 million shells, 25,000 sea mines and 2,500 bombs with explosives during the war.[8]

Explosives such at TNT are chemical compounds while other types of explosives like cordite and gunpowder are mixtures of chemicals. The development and production of explosives during the war therefore relied extensively on a knowledge and understanding of chemistry: 'The explosives industry is essentially a chemical industry and is most efficiently conducted under close chemical supervision', observed Charles E. Munroe, an explosives expert at George Washington University, USA, in a short report published in 1915.[9]

In Germany, an estimated 2,000 chemists worked in the munitions industries, while in Britain some 800–1,000 chemists worked in explosives factories and related chemical industries.[10] Munitions industries in the United States also employed several hundred chemists after the country entered the war in 1917.

Trade in explosives rose substantially during the war, the growth in exports of explosives from the United States, for example, was spectacular. The total foreign sales of explosives for 1918 'is the highest since the war started … [showing] our ability to provide for our own needs without interfering with supplies to the Allies', comments O.P. Hopkins, a reporter in Washington DC, in 1918.[11] The exports were valued at $6 million in 1914, but by 1918 had risen to $379 million.[12] Moreover, shipments to the American Expeditionary Force (AEF) in Europe were not included in the 1918 total. Explosives exported from the United States during the war included dynamite, fuses, gunpowder, and loaded cartridges, shells and projectiles. Following the end of the war, exports of explosives from the United States fell dramatically, falling to a value of £28 million in 1920.

Health and safety risks

The manufacture of explosives and munitions on such a vast scale was fraught with danger, and stringent safety precautions were put in place in munitions factories to prevent explosions, although the standards varied from country to country.[13] In Britain, for example, buttons on garments worn inside the factories were made of wood rather than metal to prevent sparks, and overalls had no pockets so workers could not bring cigarettes, pipes, matches or other items into the works. Even so, a number of devastating explosions occurred in the country's munitions factories during the First World War.

In the early afternoon of Sunday 2 April 1916, for example, about 15 tons of TNT and 150 tons of ammonium nitrate blew up at the Explosives Loading Company factory near Faversham in Kent, England.[14] The blast, which became known as 'the Great Explosion' killed over 100 people, destroyed five buildings at the factory, created a massive crater and could be heard some 100 miles away in Norwich and across the English Channel in France. It also set off a train of other explosions at the site, seriously damaged the adjacent Cotton Powder Company factory which produced guncotton, and shattered the windows of shops and houses over a wide area. The initial explosion was caused by a fire in the building where the explosives were stored, which started when some empty linen bags waiting to be filled with TNT were ignited, it was thought, by a spark from a nearby boiler house.

In January the following year, about 50 tons of TNT exploded at a chemical plant that manufactured the explosive. The plant, owned by the Brunner Mond company, was located in Silvertown, an industrial area on the north bank of the River Thames, east of London. The explosion, which once again was caused by a fire, killed seventy-three people, injured several hundred, and destroyed much of the factory and houses in the surrounding area.

The majority of employees in British factories that manufactured explosives and filled shells and bombs in the First World War were women. For example, in 1917, the munitions works at Gretna that produced cordite employed 11,576 women and 5,066 men.[15] Overall, some 75,000 women were employed in the British munitions industry in the middle of the war.[16] French munitions factories, on the other hand, tended to employ wounded or convalescent servicemen.[13]

Female workers in British munitions factories, also known as 'munitionettes', faced not just the risk of explosions but also health problems arising from exposure to toxic chemicals, most notoriously TNT. The manufacture of high explosives 'may produce symptoms not hitherto familiar to the majority of practitioners', notes a memorandum issued in 1916 by the Health of Munition Workers Committee appointed by Ministry of Munitions.[17] The memo describes the symptoms of exposure to TNT as: unusual drowsiness, frontal headache, eczema and loss of appetite, although these symptoms were generally slight at first and quickly disappeared 'if exposure ceases'. Continued exposure, however, resulted in painful limbs, shortness of breath, vomiting, anaemia, palpitations, a rapid weak pulse, jaundice, cyanosis (unusual blueness of the skin) and bile-stained urine.

A subsequent official communication published by the Ministry of Munitions in 1916 observes that 'there had been little practical experience of the toxic action of TNT' before the start of the war.[18] The communication notes that when skin or hair comes into contact with the yellow TNT powder 'a characteristic yellow or tawny orange stain' is produced that lasts for some weeks. The munitionettes who suffered from these symptoms therefore became known as 'canaries' or 'canary girls'.

Prolonged overexposure to TNT powder could lead to an unpleasant death. Hugh Thursfield, a physician at St Bartholomew's Hospital in London, provides an account of one such case of 'trinitrotoluol poisoning' in the 4 November 1916 issue of the *British Medical Journal*.[19] A strong, well-developed and muscular woman, aged 24, had been employed for four months as an examiner, inspecting bags of the yellow TNT powder, in a munitions factory before her fatal illness. The atmosphere was filled with 'enough' dust to make her sneeze she said, complaining of a bitter taste in her mouth and darker coloured urine about a month before she was taken ill. A fortnight later she lost her appetite, developed a headache and started vomiting. Then, after a few days, she noticed that she was slightly yellow, a symptom which persisted. She became deeply jaundiced and her urine was stained black with bile pigments.

After suffering from these symptoms for a further ten days, she was admitted to St Bartholomew's Hospital where she became drowsy and continued to vomit.

Over the next three days, she became restless and started shouting and moaning in a state of delirium. The following day she was 'quite unconscious and wildly delirious'. She died in the evening.

What are explosives?

Explosives are materials that pack a lot of energy into a small volume. An explosion occurs when this energy is suddenly released. The energy in military explosives such as TNT is stored in the chemical bonds that bind the atoms together in these materials. Ignition or detonation of the explosive initiates extremely fast chemical reactions that break these bonds and release the stored energy. As a result, the explosive rapidly breaks down into smaller chemical products, most notably gases such as carbon dioxide and nitrogen. At the same time, the explosion generates a lot of light and heat.

The explosives used in First World War munitions are often divided into two classes: low explosives and high explosives. Low explosives are generally powdered solids that, when ignited, burn to produce gaseous and solid products. This combustion process, which occurs at speeds less than the speed of sound, is known as 'deflagration'. When confined inside a gun barrel, the gaseous combustion products rapidly reach sufficiently high pressures to propel a bullet or shell along the barrel and out of the muzzle at high speed. For this reason, these types of explosives are known as 'propellant explosives'.

Propellant explosives are themselves divided into two classes. The first class, a typical example being gunpowder, produces not only gases but also relatively high amounts of solid products which are highly visible as smoke when used in firearms or by artillery. Conversely, smokeless powders, such as cordite, produce mainly gaseous products and relatively small amounts of solids.

High explosives are materials that are capable of decomposing and releasing the energy stored in their chemical bonds virtually instantaneously. This type of explosion, known as detonation, creates a shock wave that travels at kilometres per second, that is at speeds much higher than the speed of sound. However, many high explosives do not detonate readily. They are therefore classified as secondary explosives because they require a primary explosive to detonate them. In some cases, a bridging explosive is inserted between the primary and secondary explosives to boost the explosion of the secondary explosive. The bridging explosive is also known as a 'booster explosive' or an exploder.

Many readers interested in military history in general and the First World War in particular will have come across a variety of low and high explosives in the literature of the war. Explosives such as lyddite and ammonal are frequently mentioned in personal accounts of the war, diaries, letters, memoirs, biographies and autobiographies. Just to give one example: British Army officer Desmond Allhusen refers to lyddite in his war diary, parts of which were published in

1987.[20] In a vivid description of the Passchendaele battlefield in 1917, he writes that the mud 'rose in squashy heaps out of pools and lakes of slimy liquid, which were sometimes black, sometimes yellow from lyddite, sometimes bright green, but never the colour that water ought to be.' Such references to explosives are found not just in the non-fiction literature of the war but also in fiction. For example, in his war novel *Birdsong*, British writer Sebastian Faulks describes how two characters, Stephen Wraysford and Jack Firebrace, discover boxes of ammonal and a large wad of guncotton in a tunnel while working underground in France during the war.[21]

So what is lyddite? What is ammonal? And what is guncotton? The following sections explain what these and other common explosives used in the First World War are. But let's start with the most familiar and oldest of all military explosives, an explosive that was used in various forms and for various purposes in the First World War, namely gunpowder.

Gunpowder

Gunpowder, a low explosive, was not always called gunpowder. That's because black powder, as the explosive is also known, was discovered long before guns were invented. When and where the powder was discovered and first used is not clear, although there is some evidence that Chinese alchemists accidentally made the powder in 220 BC.[22]

Around 1,000 AD, the Chinese were known to be using gunpowder-type incendiary mixtures for fire crackers and as propellants for crude rockets.[23] The mixtures contained potassium nitrate (also known as saltpetre), sulfur and other materials. It was not until the middle of thirteenth century that Roger Bacon, an English philosopher, alchemist and Franciscan monk, introduced black powder into Europe. The gun was invented in the early fourteenth century – about three decades after Bacon's death in 1292.

The composition of gunpowder varies depending on its use. It is typically a mixture of seventy-five parts potassium nitrate, fifteen parts charcoal and ten parts sulfur. Nitrate is an oxygen-rich component of a range of explosives, containing three oxygen atoms for every nitrogen atom.

A loose mixture of the three components of gunpowder is difficult to ignite and keep alight, so a process was developed in the fifteen century to improve the burning characteristics of the powder by bringing its components into intimate contact with each other. The process involves milling a mixture of the three ingredients with a small amount of a liquid, originally alcohol or urine,[24] but subsequently water, to form a 'cake' . The cake is broken up into 'grain', dried and then granulated in a process known as 'corning'. Finally, the grain is passed over sieves of varying mesh to produce gunpowder with different grain sizes. The *Treatise on Ammunition* published by the British War Office in

1915 refers to several grades of gunpowder.[25] They range from an 'impalpable dust' of gunpowder known as 'mealed powder' to large-grain gunpowder, and 'pebble powder' which consisted of cubes of gunpowder cut from the cake. Another mixture, called 'P mixture', was a mixture of pebble powder and fine-grain powders.

When gunpowder is ignited, a complex series of chemical reactions take place. In essence, as the gunpowder burns, the nitrate, which is an oxidant, oxidises the fuel, that is the charcoal and sulfur. Potassium nitrate is normally used as the oxidant as it is easier to keep dry; sodium nitrate tends to pick up moisture from the atmosphere. The rate of oxidation or burning of the fuel depends on the size of the grains and the amount of sulfur present: the larger the grain size, the slower the burning. Similarly, less sulfur leads to slower combustion and also lowers the ignition temperature of the mixture.

The combustion process leads not only to the release of a lot of energy, in the form of heat, but also to the formation of numerous solid and gaseous chemical products, including gases such as nitrogen, carbon dioxide, and a foul-smelling and highly toxic chemical compound known hydrogen sulfide.

The use of gunpowder as a propellant for military weapons relied on the rapid generation and pressurisation of these gases within the confines of a firing chamber. However, the propellant suffered from several disadvantages. For example, as it burnt it fouled guns by depositing lots of solid particles inside the gun barrels, and it also generated volumes of smoke that revealed the position of the gunners. By the end of the nineteenth century, smokeless powders such as cordite had therefore begun to replace gunpowder as the propellant of choice.

Even so, gunpowder was still used extensively during the war for a variety of other purposes. In his memoir *Storm of Steel*, German soldier Ernst Jünger refers to the 'visionary excitement' of 'piercing fogs of gunpowder' generated by shell bursts on a day in August 1916 when German frontline trenches at the village of Guillemont were attacked by British troops during the Somme Offensive.[26]

Propellant cartridges of smokeless powders such as cordite also required a primer to initiate combustion. Fine-grain black powder that was sensitive to sparks was sometimes used for this purpose. However, the sulfur in the powder 'was considered to have a bad effect on the stability of the cordite', according to the 1915 War Office *Treatise on Ammunition*.[25] Sulfur-free gunpowder of various grain sizes, containing 70 per cent potassium nitrate and 30 per cent charcoal, was therefore used as an igniter in many types of cordite cartridge.

Gunpowder was used as a bursting charge in a variety of shells during the war. For example, British common shells were typically filled with a bursting charge of P mixture, with fine-grain gunpowder used to prime the bursting charge. Fine grain or mealed gunpowder was used in fuses and ignition tubes, and shrapnel shells employed fine-grain gunpowder as a bursting charge. A mixture of sixty parts mealed powder and forty parts nitrocellulose, known as 'composition priming', was also used to form 'an easily ignitable priming round' in some types of tube, fuse and primer.[27]

Large quantities of gunpowder were employed as mine explosives in the underground warfare carried out by specialist British, French and German tunnelling companies during the First World War. This type of warfare involved digging shafts and underground galleries that led below no-man's-land to beneath enemy lines. Mine chambers below enemy positions were then packed with explosives in order to blow massive craters in the ground above, devastate enemy positions and kill enemy troops. Other mines, known as 'camouflets', were used defensively in countermining operations that aimed to destroy enemy tunnels and entomb enemy miners.

The earliest and one of the most famous British mining operations in the First World War took place in 1915 during the Battle of Hill 60. The hill, a low ridge in the Ypres Salient, was occupied by the Germans. In March and April, a British tunnelling company dug tunnels stretching for more than 100yd under the hill. The miners then dragged bags of explosives, including ninety-four 100lb bags of gunpowder, to the mineheads, winched them down the mine shafts and shifted them through the tunnels to the mine chambers.[28] The tunnelling company exploded five mines below the German trenches on Hill 60 on 17 April. The explosion shot debris several hundred feet into the air, showered it over several hundred yards in all directions and killed some 150 Germans. At the same time, the Allies started an intense artillery bombardment of the German defences, and an assaulting party of British troops then attacked the Germans and captured the hill. Twenty Germans, stunned by the explosion, were taken prisoner, with the British suffering just seven casualties during the operation.

As the war progressed, the use of gunpowder for military mining was phased out in favour of other explosives. Ammonal, which had superior detonation properties, soon became the standard explosive for mining operations.

Brown powder

The grain size of gunpowder used as a propellant is critical: if the grains are too small, the gunpowder burns too quickly. As a result, high gas pressures rapidly build up and burst the gun before the projectile is ejected. Even gunpowder with large grain sizes, such as pebble powder, burns too fast for some types of guns, such as cannons.

A form of gunpowder called 'brown powder', so-called because of its colour, was therefore used for guns that required a slower-burning propellant. The powder, also known as 'cocoa powder' or 'slow-burning cocoa', contained less sulfur but more potassium nitrate and charcoal than black powder. However, because brown powder was difficult to ignite, it required a black powder primer to initiate its combustion.

Slow match and quickmatch fuses

Slow match was used as a safety fuse for grenades such as the Mills bomb. The slow-burning material was typically prepared by boiling a hemp cord in a solution of potassium nitrate, before the cord was then dried.

Quickmatch was a string-like fuse used for priming star shells and incendiary shells, and was made by boiling a cotton wick in a solution of mealed gunpowder and gum Arabic. Quickmatch confined in a paper pipe is known as 'piped match'.

Guncotton

Guncotton is a yellowish-white explosive chemical material that looks like raw cotton. Like so many chemicals, it has a variety of chemical names, including cellulose hexanitrate, nitrocellulose and pyroxylin. The material can be used as a high or low explosive depending on how it is prepared.

The explosive is made by treating cellulose with a mixture of concentrated nitric and sulfuric acids in a process known as nitration. The explosive nature of guncotton arises from the relatively high levels of nitrate in the material. The fuel component of guncotton is cellulose, which is a naturally-occurring organic polymer consisting of glucose units strung together in long chains. It is a constituent of cotton and other plants, is insoluble in water and is the most abundant natural organic polymer on earth. Nitrocelluloses with lower levels of nitrate are flammable, but not explosive, soluble in some organic solvents and are used to make plastics such as celluloid, lacquers, inks and adhesives.

Guncotton was discovered by German-Swiss chemist Christian Friedrich Schönbein (1799–1868) in 1846, who prepared it from cotton wool and a mixture of the two acids. The explosive burnt without producing vast quantities of smoke and was much more powerful than gunpowder. However, it proved difficult to manufacture safely and it burnt far too quickly for use as a propellant.

In 1884, French chemist Paul Vieille (1854–1934) showed that a safe gelatinous version of the smokeless powder, which became known as '*Poudre B*', could be prepared by mixing nitrated cellulose with a mixture of two organic solvents: ethyl ether and ethanol. Another organic chemical, amyl alcohol, was added to the mixture to scavenge any acid remaining in the nitrocellulose and thereby stabilise the gel. After further processing, a plastic form of the powder was produced that could be rolled into sheets and cut into flakes or extruded into cords, strips or tubes and cut to the desired length. Within a few years of the discovery, the French military adopted the powder for use as a smokeless propellant for firearms and artillery, with other countries following suit. In 1910, the French authorities decided to replace amyl alcohol with the nitrogen-containing organic chemical diphenylamine as the stabiliser.

The supply of cotton cellulose for the production of guncotton proved a problem during the First World War, especially for the Germans who relied on imported cotton for its manufacture. The British naval blockade had put a stop to the supply and the Germans therefore turned to wood pulp as a possible source of cellulose.[29] The process was based on a discovery by Edward Schultze, a Prussian artillery officer, in 1846. In the 1860s he showed that nitrocellulose could be prepared by nitrating small pieces of wood with nitric acid and a mixture of barium nitrate and potassium nitrate. Following Schönbein's discovery, numerous attempts were made to exploit other sources of cellulose for the production of nitrocellulose explosives and propellants. The sources included grass, flax, solid fruits, nuts and their shells, jute fibre, maize, hemp, gorse, bamboo, cornstalks and corncob.[30] None of these proved successful owing to the difficulties in purifying the cellulose or obtaining sufficient quantities of the material to be useful.

Guncotton was not only the standard component of the smokeless powder propellants used in the First World War. It was also employed for other purposes, such as in torpedo warheads and some types of grenade, and as a detonating explosive in priming devices and vent-sealing tubes used to fire charges of propellant in guns. Guncotton dust, mixed with mealed gunpowder, was used in some types of tube.

The superior explosive power of guncotton compared with gunpowder made it particularly useful for blasting; for example, for blowing up bridges and in some early mining operations in the First World War. On 19 July 1915, for instance, a mine containing almost 5,000lb of gunpowder, guncotton and ammonal was detonated by the British at Hooge in Flanders, Belgium. The explosion left a crater some 120ft in diameter and 20ft deep.

Guncotton, however, was less powerful and not as safe as ammonal. In the absence of additional nitrate or some other oxidising agent, the explosive does not contain sufficient oxygen to convert all of its carbon to carbon dioxide and all of its hydrogen to water when it explodes. Incomplete oxidation of guncotton produces a mixture of gases including hydrogen and carbon monoxide. Neither gas can be seen or smelled and both are highly flammable, plus carbon monoxide is highly toxic. The underground explosion of a mine containing guncotton, particularly a camouflet, was therefore extremely dangerous as the toxic and flammable gases could blow back through the tunnels. According to one report, one British tunnelling company suffered 150 cases of carbon monoxide poisoning, including sixteen fatalities, over a six week period in 1915.[31]

Tonite was also a safer blasting explosive than guncotton and was prepared by mixing equal weights of guncotton and the inorganic compound barium nitrate. The barium nitrate provided additional oxygen to ensure that all the carbon and hydrogen in the guncotton was converted to carbon dioxide and water respectively when it exploded. The explosive was used by the British as a bursting charge in some types of rifle grenades and stick grenades, and also as a blasting explosive in some mining operations.

Nitroglycerine

Nitroglycerine, spelt nitroglycerin in the United States, was a component of a variety of explosives used during the war. The compound, also known as glycerol trinitrate, nitroglycerol and by many other names, is a pale yellow oily liquid. Like guncotton, nitroglycerine derives its explosive power from the high proportion of nitrate in the compound, each molecule of the compound containing three carbon atoms, five hydrogen atoms and three nitrate groups.

The compound was discovered by Italian organic chemist Ascanio Sobrero (1818–1888) in 1846.[32] It is produced by slowly adding pure glycerine to a mixture of concentrated nitric and sulfuric acids at a temperature just above the freezing point of water (0°C). Glycerine is an oily organic compound obtained as a by-product of soap manufacture, although additional supplies were obtained during the war using a process that involved fermenting sugar.[33]

Pure nitroglycerine is unstable and detonates when given a sharp blow or when it is rapidly heated. It is therefore far too dangerous to use by itself and for this reason is incorporated into other materials to make it safe to handle and transport. During the First World War it was used to manufacture dynamite, cordite and many other explosives.

Dynamite

In 1866, Swedish chemist and industrialist Alfred Nobel (1833–1896) discovered that nitroglycerine could be made safe to handle by using diatomaceous earth, an inert naturally-occurring porous solid, to absorb the oily liquid. The earth, known as diatomite, kieselguhr or kieselgur, consists of the fossilised remains of a group of algae known as diatoms, which contain silicon dioxide. Nobel patented the mixture under the name 'dynamite' in 1867.

The high-explosive mixture proved particularly useful as a blasting explosive for the demolition of buildings, mining and quarrying. It could readily be packed, stored, transported and, with a detonator in a blasting cap to set it off, safely exploded.

Nobel made a fortune from his patented inventions of dynamite and other explosives. In his will, he left the bulk of this fortune to establish the Nobel Foundation in Stockholm. The foundation provides the funds for the Nobel Prizes in Chemistry, Physics, Physiology or Medicine, Literature and Peace, which have been awarded annually since 1901, and the Prize in Economic Sciences in Memory of Alfred Nobel which has been awarded every year since 1969.

The dynamite patented by Nobel in 1867 contained about 75 per cent nitroglycerine and 25 per cent diatomaceous earth. Since then, dynamite explosives with other absorbent materials have also been developed. In the First World War, for example, a type of porous clay known as 'fuller's earth' was used as the absorbent material in some forms of dynamite.[34]

A notable example of the use of dynamite in the First World War occurred during the British Royal Navy raid on the German-held Belgian port of Zeebrugge on the night of 22/23 April 1918. Part of the raid involved landing demolition parties on the Zeebrugge Mole. The mole, a sturdy breakwater with a railway line, buildings, hangars and store-sheds, protruded in a curve from the port into the sea and guarded the entrance to the Bruges Canal. The demolition parties laid and set off dynamite charges, destroying or damaging many of its buildings.

Gelignite

After his discovery of dynamite, Nobel continued to experiment with the explosive. He was particularly concerned that the highly unstable nitroglycerine occasionally exuded as an oily liquid from the kieselguhr absorbent in a process known as sweating. In 1875, Nobel showed that a blasting explosive that did not sweat could be made using a mixture of nitroglycerine and a small amount of nitrocellulose dissolved in a mixture of ethanol and ether. The nitrocellulose in the solution was not so highly nitrated as guncotton and was called collodion. The explosive mixture of collodion and nitroglycerine had the texture of jelly and was water resistant, becoming known as gelignite or blasting gelatine. The explosive was not only safer than dynamite but also much more powerful.

Ballistite

Nobel subsequently showed that a mixture of nitroglycerine and collodion in equal proportions could be combined with a small percentage of camphor to make a smokeless propellant. He patented the mixture under the name 'ballistite' in 1887. Camphor is an organic chemical compound that occurs in camphor laurel trees and some other laurel species. When pure, camphor is a waxy white or translucent solid. The compound stabilised the propellant by removing any acid formed by decomposition of the explosives.

Ballistite can be cut into flakes and it burns vigorously. In the First World War, it was used by the British as a smokeless propellant in the earliest cartridges for QF 15-pounder guns and 2.95in BL howitzers.[27] As the propellant contains two explosives, nitroglycerine and nitrocellulose, it is known as a double-based propellant, whereas propellants based on nitrocellulose only are known as single-based propellants.

Cordite

Between 1914 and 1918, the British Army and British Royal Navy fired almost 250 million shells. Cordite was the standard, although not the only propellant used in cartridges to fire these shells. Cordite was also used in ammunition for small arms; for example, most versions of the Webley revolver, the standard issue service pistol, fired a cartridge, the .455 Webley, that contained cordite as the propellant.

Like ballistite, cordite was a double-based smokeless propellant containing both nitrocellulose and nitroglycerine. But unlike ballistite, cordite contained petroleum jelly rather than camphor to stabilise the explosive mixture. Furthermore, the jelly, also known as mineral jelly or by its brand name Vaseline, waterproofed the explosive mixture.

Cordite was patented in 1889 jointly by English chemist and explosives expert Frederick Abel (1827–1902) and Scottish chemist and physicist James Dewar (1842–1923). Many years earlier while working as a chemist in the British War Department, Abel had developed a new and safer method for making guncotton. The process involved thoroughly cleaning and drying the cotton to avoid spontaneous explosion of the nitrated product. Dewar, a professor at Cambridge University and the Royal Institution, London, is famous for inventing the vacuum flask, otherwise called the Dewar flask or thermos flask, in 1892.

Abel and Dewar's smokeless propellant consisted of 58 per cent nitroglycerine, 37 per cent guncotton and 5 per cent petroleum jelly. The explosive was made by mixing the two explosives together to form a paste, before the organic solvent acetone, or propanone as it is sometimes called, and then the petroleum were added to the paste. The resultant dough was extruded from a hydraulic press in the form of a cord, hence the name cordite, and was finally dried to remove most of the acetone. The product was then typically cut into sticks of the required length and tied in bundles for insertion into cartridges.

The similarity in the composition of cordite and ballistite prompted Nobel to sue Abel and Dewar for patent infringement. Nobel lost his case because cordite contained guncotton, an insoluble form of nitrocellulose, whereas ballistite was prepared using a soluble form of the material, namely collodion.

The type of cordite invented by Abel and Dewar became known as Cordite Mark I. However, the relatively high proportion of nitroglycerine in the propellant caused severe erosion in the gun barrel bores. To overcome the problem, the proportions of nitroglycerine and guncotton were reduced to 30 per cent and increased to 65 per cent respectively, while the proportion of petroleum jelly was kept at 5 per cent. This version of the propellant was called Cordite MD, where MD is an abbreviation of MoDified.

His Majesty's Factory, Gretna, which began producing cordite in April 1916, was the largest explosives factories in the British Empire. The factory stretched for miles, employed over 15,000 people and had the capacity to produce 1,000 tons of cordite per week. It was at this factory that cordite paste became famously dubbed 'devil's porridge' following a then secret visit by Scottish writer Sir Arthur Conan Doyle (1859–1930) during the war. He observed that the workers kneaded the nitroglycerine and guncotton into 'a sort of devil's porridge'.

A variant known as Cordite RDB, which stands for Research Department formula B, was developed at the Research Department of the Royal Arsenal at Woolwich in south-east London. This type of cordite replaced insoluble guncotton with the soluble nitrocellulose called collodion. The exact composition was 52 per cent collodion, 42 per cent nitroglycerine and 6 per cent petroleum jelly. A mixture of ethanol and ether was used as the solvent in place of acetone which was in short supply during the war.

The manufacture, storage and use of cordite were not without their dangers, however. One of the most spectacular examples was the sinking of the British Royal Navy battlecruiser HMS *Indefatigable* in the North Sea on 31 May 1916 during the Battle of Jutland between the German Navy's High Seas Fleet and the British Royal Navy's Grand Fleet. The *Indefatigable*'s armament included eight 12in breech-loading guns. Each shell fired by these guns required four 65lb cordite cartridges. Less than two hours after the start of the battle, several large-calibre shells from the German battlecruiser SMS *Von der Tann* hit the *Indefatigable*. Within minutes, the magazines containing bagged cordite cartridges blew up and the ship sank almost immediately. Only three members of the crew of 1,017 survived. Soon after, on the same day, the British battlecruisers HMS *Queen Mary* and HMS *Invincible* also suffered similar fates with most of their crews perishing.

In pursuit of acetone

Acetone is a common and versatile organic solvent, and was required in copious amounts for the manufacture of Cordite MD and the dope used to tighten airplane wings. Nowadays, it is a familiar component of nail polish remover. During the First World War, wood was used as source material for acetone. It was heated to produce a dark liquid called pyroligneous acid or wood vinegar. The process is known as pyrolysis or destructive distillation. Acetic acid in the liquid was then converted to calcium acetate by neutralising it with calcium oxide, commonly known as lime. Finally, the calcium acetate, a white solid compound, was heated to form acetone and calcium carbonate as a by-product. About 100 tons of wood were needed to produce one ton of acetone. Before the war, much of the wood used in Britain for this process was imported from timber-growing countries such as the United States and Austria. However, the supply of acetone from wood during the war proved insufficient to meet British demands so alternative sources were sought.

The fermentation of starch obtained from grain and other sources provided one solution. Fermentation is a biochemical process that relies on microorganisms such as yeast or certain bacteria to act on sugars to produce ethanol and other useful chemical products. Fermentation using yeast forms the basis of bread, beer and wine production.

In 1912, a chemistry lecturer at the University of Manchester who had been carrying out research on fermentation discovered a type of bacterium that was able to produce large quantities of acetone from the starch in grain. The chemist was Chaim Weizmann (1874–1952) and the name of the bacterium was *Clostridium acetobutylicum*, also known as the 'Weizmann organism'. The process, which Weizmann patented, became known as ABE fermentation as it produced a mixture of acetone (A), a relative of ethanol known as butanol (B), and ethanol itself (E).

Weizmann was born in Belarus, then part of the Russian Empire, and studied chemistry in Germany and Switzerland before lecturing on chemistry at the University of Geneva. He emigrated to England in 1904 and subsequently became a naturalised British subject. Following his move to Manchester, Weizmann developed and patented his fermentation process. In 1915 he started to work with the British Admiralty and showed that 12 tons of acetone could be produced from 100 tons of maize using his process. The Admiralty subsequently constructed an acetone plant with fermentation vessels at the Royal Navy Cordite Factory at Holton Heath, Dorset. By 1917, Weizmann's process was producing acetone at a rate of almost 3,000 tons annually at the Holten Heath plant and other sites in Britain.

The success of the process relied on a steady supply of cereals as a source of starch. However, much of the grain, particularly maize, was imported and was also needed for food. By 1917, with the growing threat of the German submarine blockade, alternative non-edible sources of starch were sought, including horse chestnuts or conkers as they are popularly known. Weizmann modified his process to produce acetone from this source, and during this period the Ministry of Munitions in Britain encouraged children to contribute to the war by collecting conkers for acetone production. Weizmann became known as the father of industrial fermentation, although he is more famous for his political activities. In 1917, he helped Arthur Balfour (1848–1930), who was Foreign Secretary in David Lloyd George's coalition government, to secure the Balfour Declaration that recommended the establishment of a national home for the Jewish people in Palestine. After the war, Weizmann became president of the World Zionist Organisation and, in 1949, became the first president of the State of Israel.

Picric acid and lyddite

Picric acid is a pale yellow crystalline solid that explodes above 300°C. Like many other explosives, guncotton and nitroglycerine for example, picric acid is a highly nitrated compound, as its chemical name 2, 4, 6-trinitrophenol indicates.

Picric acid, its preparation and its explosive properties were well known in the first half of the nineteenth century. Then, in 1885, French chemist and explosives expert François Eugène Turpin patented the use of picric acid that had been melted and allowed to solidify as a blasting charge or a bursting charge for artillery shells. A small amount of another organic compound, such as camphor, collodion or TNT was usually added to the picric acid in order to lower its melting point and allow it to be poured more easily into shells.

The compound, widely used as a high explosive in the war, was prepared in two steps. First, phenol, a chemical obtained from coal tar, was treated with concentrated sulfuric acid in a process known as sulfonation. The sulfonated product was then nitrated with concentrated nitric acid. In 1914, Georges Darzens (1867–1954), a Russian-born chemist who was a professor at École Polytechnique in Paris, France, devised an alternative process for manufacturing the explosive and showed that picric acid could be made from a nitrogen-containing organic compound known as aniline.[35]

Picric acid was the principle component of the French high explosive 'melinite'. The British equivalent 'lyddite' was named after Lydd, the town in Kent where it was manufactured. The dark yellow material was used as a bursting charge in shells fired by the British Army in the Second Boer War (1899–1902), fought between the British and Boers (also known as Afrikaners) for the conquest of southern Africa. The common high-explosive and armour-piercing shells used by the British early on in the First World War contained lyddite.

Ammonium picrate, picric powder, dunnite

Ammonium picrate, a yellow crystalline material made by treating picric acid with ammonia, is also explosive. In the First World War, the explosive was mixed with saltpetre, also known as potassium nitrate, to make picric powder, a powder that was used as an exploder to boost the detonation of lyddite in shells.

The United States Navy employed ammonium picrate as a bursting charge in armour-piercing shells. The explosive, known as 'Explosive D' or 'dunnite' after Colonel Beverly W. Dunn of the United States Army who developed it, was less powerful than lyddite.

Trinitrotoluene

Trinitrotoluene is a high explosive and, as its name indicates, is a highly nitrated organic compound. The compound, a yellow crystalline material, is also known as TNT, trotyl, trinitrotoluol or 2, 4, 6-trinitrotoluene. It was widely used by various armies in the First World War, either by itself or as a component of explosive mixtures, as the exploder or bursting charge in high-explosive artillery shells, and as a blasting explosive.

The compound was first prepared by German chemist Julius Wilbrand (1839–1906) in 1863. Germany began manufacturing the material in 1891 and in 1902 the German Army adopted it as a replacement for picric acid (lyddite) in artillery shells. Five years later, Britain began to replace lyddite with TNT as a filling for high-explosive shells. Both high explosives were used during the First World War, although the French preferred picric acid as it was a more powerful explosive than TNT. Even so, TNT had several advantages over lyddite. First, the acidic nature of picric acid meant that in the presence of moisture it could react with and corrode shell cases by forming metallic picrate salts. TNT, on the other hand, is not an acid, does not pick up moisture and does not react with metals. Second, pure TNT melts at 80°C compared with 122°C for picric acid, which meant that TNT could be melted and poured into artillery shells at a lower temperature. Third, TNT was more stable than lyddite, was insensitive to shock and required a powerful detonator to explode: 'A rifle bullet fired through a case of TNT has no explosive effect' observed an explosives expert at the Carnegie Institute of Technology, Pittsburgh, Philadephia in 1917.[36] TNT was therefore safer to handle, store and transport than lyddite, could be used in a less than pure form and was also less expensive to produce.

TNT was manufactured by nitrating toluene, an organic solvent obtained principally by distilling coal tar and sometimes petroleum, using a mixture of nitric and sulfuric acids. Approximately 2 tons of toluene was needed for every 3 tons of TNT produced. During the war, Britain typically needed around 600,000 tons of coal to produce the 720 tons of toluene required to produce its weekly output of TNT.[7]

Blastine and dinitrotoluene

Guncotton not only gave off large quantities of carbon monoxide when it exploded, but it also had 'inferior cratering properties', according to British tunneller Lieutenant F. J. Mulqueen of the 172th Tunnelling Company, Royal Engineers, whereas Blastine 'was quick acting and more effective'.[37] Blastine was one of a number of explosives used in the First World War that contained the explosive compound dinitrotoluene (DNT).[38] The yellow crystalline compound has a similar chemical structure to TNT but contains less nitrogen and oxygen.

During the war, the French and Italians employed smokeless powders containing nitrocellulose, nitroglycerine and DNT, which acted as an anti-flash additive. Cordite and similar smokeless powders containing nitrocellulose produced a bright flash from the muzzle of a gun when it was fired. The flash, which could reveal the position of the gun to the enemy, was caused by the ignition of a cloud of hot combustible propellant gases as it emerged from the muzzle and mixed with the oxygen in the air. These flammable gases, as we saw above, were produced by the incomplete oxidation of nitrocellulose.

However, they only caught fire if the temperature inside the barrel of the gun was sufficiently high. Flash suppressants such as DNT lowered the temperature at which the propellant powder burnt, so the temperature inside the barrel was therefore not high enough to ignite the gases. Smokeless powders that do not produce a flash when a gun is fired are known as 'flashless' powders.

Ammonium nitrate

Ammonium nitrate, a white solid, is the ammonium salt of nitric acid. The compound consists of three chemical elements: nitrogen, hydrogen and oxygen. When detonated or heated to a high temperature under close confinement, the compound decomposes explosively.

While explosive chemicals such as nitroglycerine and TNT are organic compounds because they contain carbon, as well as nitrogen and oxygen, ammonium nitrate is an inorganic compound as it does not contain carbon. However, because it contains a high proportion of oxygen, it supports the combustion of carbon-containing materials. Moreover, it explodes readily when contaminated by combustible materials.

Like many other explosive chemicals used in the Great War, the explosive properties of ammonium nitrate were well known before the start of the war. The salt was first prepared by German-Dutch chemist Johann Glauber (1604–1670) in 1654. At the beginning of the nineteenth century it was considered as a possible replacement for potassium nitrate in gunpowder and by the middle of the century there was at least one report of a mixture of charcoal and ammonium nitrate powder exploding when heated.[39]

Ammonium nitrate was a component of explosive mixtures such as amatol and ammonal used in the war. And because the salt has a high nitrogen content and is soluble in water, it was and still is used as a fertiliser by the agricultural industry. Nowadays, mixtures known as ANFO (Ammonium Nitrate Fuel Oil) explosives are widely used as blasting explosives in mining and quarrying. They contain a mixture of ammonium nitrate and a carbon-containing fuel such as heating oil. Terrorists have used such mixtures in improvised bombs, also known as fertiliser bombs.

Amatol

As the First World War progressed, the demand for the high explosive TNT began to outstrip supply. Production of TNT relied heavily on a reliable supply of toluene, a chemical that was also needed to manufacture dyes and other synthetic organic chemicals.

Before the war, Britain, like other countries around the world, had imported dyes and other synthetic organic chemicals from Germany. With the outbreak of war, Britain had to start making its own dyes and chemicals to replace the imports and, as a result, the production of toluene from coal tar and petroleum could not keep up with demand and the chemical became increasingly scarce. Similarly, in Germany, which had imported coal tar from Britain before the war, supplies of chemicals such as toluene began to dry up.

To help conserve supplies of both toluene and TNT, both Germany and Britain in 1915 started to fill shells with a high explosive consisting of a mixture of 40 per cent ammonium nitrate and 60 per cent TNT. The explosive, known in Britain as 'amatol', was prepared by adding ammonium nitrate powder to molten TNT, pouring the molten mixture into shells and then allowing it to solidify. The mixture had similar explosive properties to pure TNT but was much safer to handle. It was also less expensive to produce because ammonium nitrate could be manufactured easily and at relatively low cost in large quantities. In Britain, the Woolwich Research Department subsequently showed that an equally effective but much cheaper version of amatol could be made by increasing the proportion of ammonium nitrate to 80 per cent and reducing that of TNT to 20 per cent. In October 1917, the United States Army also adopted amatol as a bursting charge for its high-explosive shells.

At the end of the war, Lord John Fletcher Moulton (1844–1921), a lawyer who had spearheaded the development and manufacture of high explosives in Britain during the war, commented that amatol 'was the greatest single thing of importance' in boosting the supply of munitions and enabling 'our armies to expend shell on an unlimited scale'.[40]

Ammonal

Ammonal was used in the First World War as a filling for hand grenades, shells and trench mortar bombs, and as a blasting explosive for mines. The British used the explosive to blow up hundreds of mines in tunnels below German lines in France and Belgium. Ammonal is an abbreviation of the names of two chemicals: AMMONium nitrate and ALuminium. It was the name given to a variety of high explosives during the war that contained both these materials. As we saw above, ammonium nitrate explodes at high temperatures, while aluminium in its powdered form is easily ignited and burns fiercely. However, ammonal is a relatively safe explosive as it is difficult to set off accidentally. For example, the material does not explode when hit by a rifle bullet, but in a confined space, detonation of the mixture results in a powerful explosion.

One type of ammonal, used by the Austrians in the war, consisted of a mixture of ammonium nitrate, aluminium powder, picric acid, TNT and carbon.[41] The Russian and Austrian armies were also reported to have used a mixture, also known as ammonal, consisting of ammonium nitrate, 12–15 per cent TNT, a small amount of powdered aluminium and a trace of charcoal.[42] A German version of ammonal, used early in the war, contained 54 per cent ammonium nitrate, 30 per cent TNT and 16 per cent flakes of aluminium.[43] As the war progressed and TNT became scarcer, the proportion of ammonium nitrate was increased to 72 per cent, and that of TNT reduced to 12 per cent.

Ammonal is about three times more powerful than gunpowder and became the standard explosive charge for British mining operations along the Western Front. According to Lieutenant F.J. Mulqueen of the 172th Tunnelling Company, 'it was easy to handle, was quick acting and generally met our requirements'.[37]

Seventeen British mines charged with ammonal were blown up on the first day of the Somme Offensive, 1 July 1916, to create breaches in the enemy lines. For example, at La Boisselle, a village just outside the town of Albert, the 179th Company of the Royal Engineers exploded two mines, one packed with a 13-ton charge of the explosive and the other with 11 tons. The mines were located 60ft apart in a Y-shaped tunnel dug by the 185th Tunnelling Company of the Royal Engineers and the explosion created a 90ft-deep crater, some 300ft in diameter. The crater, which still exists today, was named the Lochnagar Crater after the trench where the main tunnel began.

Prior to the infantry assault by the Allies on Messines Ridge, near Ypres, on 7 June 1917, army engineers from Australia, Britain, Canada and New Zealand dug, over a twelve-month period, 4 miles of tunnels beneath a 10-mile section of the German frontline. They laid twenty-two mines packed with a total of around 450 tons of ammonal. The Germans conducted counter-mining operations and detonated one of the mines. Of the twenty-one remaining mines, nineteen exploded, killing or burying alive around 10,000 German troops and wrecking German positions as well as the town of Messines.

The explosions, together with a ten-day bombardment that had preceded the assault, paved the way for the infantry to break through and capture the ridge. The Messines Ridge offensive provides a prime example of how the manufacture of a variety of chemicals – ammonium nitrate, TNT and aluminium in this case – on an immense scale had a major impact on the course of the war.

Schneiderite

Schneiderite was a high explosive used by the French, Italians and Russians in the First World War as a bursting charge in small- and medium-size shells. The explosive consisted of seven parts ammonium nitrate and one part dinitronaphthalene. The latter chemical is a nitrated version of naphthalene, a flammable organic compound that once was used to make mothballs. The explosive was prepared by grinding the two solid components together in a powder mill, before the powder was then compressed into the shells.

Mercury fulminate

Mercury fulminate is a chemical compound containing the metallic element mercury and the non-metallic elements carbon, nitrogen and oxygen. The chemical is a primary explosive that detonates spontaneously and violently by shock, friction or when heated or hit by a spark. The explosive was therefore stored under water before small quantities were then carefully dried for use.

The chemical was employed in the First World War, either by itself or mixed with other substances, as the primer in pistol and rifle cartridges. It was also used in caps and tubes to fire the propellant charges in guns and howitzers, and in fuses to detonate the bursting charges in shells and hand grenades.

Like other explosive chemical compounds used during the war, mercury fulminate was not a new discovery. It was first prepared in the seventeenth century by German chemist Johann Kunckel (1630–1703). The discovery was largely ignored until 1800 when British chemist Edward Howard (1774–1816) described its preparation and properties in a paper presented to the Royal Society in London.[44] Howard demonstrated that the grey crystalline compound was produced by treating a solution of mercury in nitric acid with ethanol. Furthermore, he showed that the product, when dry, could be detonated by an electric spark or by striking it with a hammer.

Howard also proposed that mercury fulminate might be used as a percussion initiator for gunpowder. However, he was unable to demonstrate its use as a percussion primer and it was not until 1807 that Scottish clergyman Alexander Forsyth (1769–1843) invented and patented a percussion cap containing a priming composition that included mercury fulminate. The composition ignited when struck by a hammer. The invention led to the demise of flintlock weapons, which employed a trigger to strike a hammer against a flint, generating a spark that ignited the gunpowder charge in the barrel.

Potassium chlorate

The priming composition employed by Forsyth in his percussion cap contained not only mercury fulminate, but also potassium chlorate, sulfur, and charcoal. Potassium chlorate, which used to be called 'chlorate of potash', is a white crystalline inorganic compound made up of the elements potassium, chlorine and oxygen. Like mercury fulminate, it is a primary explosive that is highly sensitive to friction and shock.

Potassium chlorate provided a source of oxygen in many of the detonating compositions used in the First World War. These primers also contained other materials such as mercury fulminate and antimony trisulfide. The explosive was also mixed with wax to make a plastic explosive for filling grenades and other munitions and is also known as a 'putty' explosive because it can be moulded by hand.

Potassium chlorate was discovered by French chemist Claude Berthollet (1748–1822). It is made by the reaction of its cousin, sodium chlorate, with potassium chloride. However, like so many other chemical explosives, the presence of impurities in potassium chlorate can have a pronounced impact on its performance.

Impurities posed a particular problem in the United States when it entered the war in 1917. Until then, potassium chlorate had been imported from abroad but the war cut off supplies and, initially, the potassium chlorate primers manufactured in the country proved to be defective. The main culprit was a compound related to potassium chlorate known as potassium bromate. It contained the elements potassium, bromine and oxygen, and was generated from bromide impurities in the potassium chloride from which potassium chlorate was made: 'The chlorate made in the United States carries a higher percentage of impurities', notes H.C. Pritham, chief chemist at the Frankford Arsenal in Philadephia, Pennsylvania, in a report published in 1917.[45] The bromate was not eliminated during the manufacturing process and as a result the primer slowly deteriorated. However, a new process was later developed in order to remove the impurity and increase the effectiveness of the potassium chlorate.

Tetryl

Tetryl is a high explosive that was also known by the British in the First World War as 'composition exploding'. The explosive was sometimes employed as a secondary detonating material, a booster explosive, in fuse magazines. It was also used in large quantities as a booster explosive in some types of high-explosive shell.

The explosive, a yellow crystalline material, is a highly nitrated organic compound. It has a variety of lengthy chemical names which provide information about the chemical composition of the compound, including 2, 4, 6-trinitrophenylmethylnitramine, a name that is sometimes abbreviated to 'nitramine'. However, this is incorrect as the term 'nitramine' in chemistry strictly refers to a group of nitrogen- and oxygen-containing organic compounds related to ammonia.

Tetryl was first prepared and studied in the 1870s and '80s, and during the war was manufactured by the nitration of the related organic compound dimethylaniline with sulfuric and nitric acid.

Nitroguanidine

Tetryl was not the only nitramine explosive used in the First World War. Another was nitroguanidine, which was made by nitrating guanidine, a chemical compound with a high nitrogen content. Guanidine was manufactured from guanine, a naturally-occurring chemical that is found in guano, the excrement of certain types of bat and seabird. Guanine also has another claim to fame as the chemical component of the genetic material DNA (an abbreviation of deoxyribonucleic acid) that occurs not just in ourselves but in all living organisms.

Like tetryl, nitroguanidine was first prepared in the 1870s. During the war, the German Army used a mixture of nitroguanidine, ammonium nitrate and paraffin as a bursting charge for trench mortar shells. They also used it mixed with nitrocellulose as a smokeless propellant.

Lead Azide

Lead azide is powerful primary explosive that explodes when struck. The compound, a white powder consisting of the chemical elements lead and nitrogen in the ratio one to six, was first brought into use by the explosives industry in France and Germany a few years before the start of the First World War. It was used in various types of detonator by the military throughout Europe during the war. For example, it was used with a booster explosive such as tetryl in the detonators of some artillery fuses.

Ophorite

Ophorite, sometimes known as magnesium ophorite, is a mixture of potassium perchlorate and magnesium.[46] The explosive was used by the Americans as a bursting charge for its 4in Stokes mortar gas, smoke and incendiary shells and by the British as an igniting and expelling charge for its 18-pounder thermite incendiary shells.[47] Thermite is a mixture of finely divided aluminium and iron oxide which, when it burns, generates an enormous amount of heat.

The perchlorate component of potassium perchlorate contains the elements chlorine and oxygen in a ratio of one to four, whereas that in chlorate is one to three. Perchlorate therefore contains more oxygen than chlorate and consequently is an excellent oxidant. When ignited in the presence of the oxidant, magnesium burns fiercely at a sufficiently high temperature to ignite thermite.

5

The Metals of War

How the war depended on a precious metal

'Platinum plays such a vital part in the war that it is nothing short of a crime to allow its use for purposes of personal adornment.' So wrote G. Shaw Scott, secretary of the British Institute of Metals, in a letter to Charles L. Parsons, secretary of the American Chemical Society (ACS), published in August 1917.[1] The purpose of the letter was to invite cooperation between the two organisations in their efforts to conserve platinum.

Platinum is one of a group of chemical elements known as precious metals; others include gold and silver. They are 'precious' because of their rarity. So how did this silvery-white precious metal make such an important contribution to the war effort? The answer lies in the manufacture of sulfuric acid and nitric acid, which were essential for the production of explosives such as guncotton, nitroglycerine and trinitrotoluene.

Much of the sulfuric acid needed for the war was manufactured by what is known as the 'contact process'. The first part of this process involves burning sulfur, or brimstone as it is also called, in air. Alternatively, a sulfur-containing mineral such as iron pyrite, sometimes known simply as pyrite or by its chemical name iron sulfide, is roasted in air. The burning of the sulfur or roasting of the sulfide mineral yields sulfur dioxide, a gaseous inorganic compound containing one part sulfur and two parts oxygen.

In the second part of the process, sulfur dioxide and the oxygen in the air react together to form sulfur trioxide, a compound that contains sulfur and oxygen in the proportions one to three. In the final step, sulfur trioxide and water combine to form sulfuric acid. As the second part of the process is slow, a catalyst is employed to speed up the reaction. The original contact process, patented by English vinegar maker Peregrine Phillips in 1831, employed finely divided platinum as a catalyst. The sulfur dioxide and air were passed through a platinum tube containing the finely divided metal.[2] By the beginning of the First World War, however, the process had changed somewhat. The sulfur trioxide was typically manufactured by passing sulfur dioxide and air over trays containing platinum deposited on granules of the inorganic chemical compound magnesium sulfate.[3] After the war, the platinum catalyst was gradually replaced by a less expensive and more effective catalyst: vanadium pentoxide, an inorganic compound consisting of one part of the metal vanadium and five parts oxygen.[4]

Sulfuric acid was also manufactured by the so-called 'lead chamber' process. The process, which dates back to the eighteenth century, employed lead-lined chambers to convert sulfur dioxide into the acid. Although the process did not involve the use of platinum or the formation of sulfur trioxide, it was less economic than the contact process. Nevertheless, it continued to be used well into the 1950s. Nowadays, virtually all sulfuric acid is made by the contact process using the vanadium pentoxide catalyst.

Sulfuric acid manufactured by the contact and lead chamber processes was required to manufacture the nitric acid needed to make explosives. Many people will remember chemistry teachers at school demonstrating how to make nitric acid from sodium nitrate. This nitrate salt, also known as Chile saltpetre, occurs naturally in abundance in Chile and Peru as a mineral in ores known as caliche. The school demonstration involved heating a mixture of the nitrate salt and concentrated sulfuric acid in a glass retort, and the same chemical procedure was used on an industrial scale to produce nitric acid from the mineral.[5]

Chile saltpetre was imported in huge quantities by countries such as Germany, Britain and the United States before the First World War, and was used not just for manufacturing nitric acid, but also as a fertiliser. In 1913 for example, Germany imported 750,000 tons of the salt from Chilean mines.[6] When war broke out in August 1914, Germany had a stockpile of half a million tons of the salt. The British Royal Navy blockade cut off further supplies and the stockpiles proved insufficient to meet the demands for nitric acid and explosives required by the German munitions industries. Germany therefore turned to ammonia, a compound that consists of nitrogen and hydrogen, as a source of the nitrogen for the manufacture of nitric acid. Crude ammonia was commonly obtained as a by-product in the production of coke and coal gas from coal. However, impurities in the by-product reduced its effectiveness for the manufacture of nitric acid. A much purer form of the compound could be produced synthetically on an industrial scale by the reaction of nitrogen extracted from air and hydrogen, an element obtained by passing steam over red-hot coke. The industrial synthesis of ammonia by this method is known as the Haber-Bosch process.

'Without such a process Germany could not have made the nitric acid required for her explosives programme, nor obtained fertilisers for food production after our blockade, and it is probably that she could not have continued the war after 1916,' observed British chemist Harold Hartley (1878–1972) in 1919.[7] Hartley, a lecturer in chemistry at Oxford University, served in the British Army and by the end of the war had been promoted to the rank of brigadier general and Controller of the Chemical Warfare Department at the British Ministry of Munitions.

The pure ammonia produced by the Haber-Bosch process was converted into nitric acid by its reaction with oxygen in the air, a process, patented in 1902, which was developed by the German chemist Wilhelm Ostwald.[8] As with the contact process for the synthesis of sulfuric acid, the Ostwald process employed platinum as a catalyst. In the later years of the war, the Germans used the process to supply all the nitric acid needed for explosives, and so the German war effort beyond 1916 relied not just on the Haber-Bosch process, as Hartley had pointed out, but also on the platinum-catalysed Ostwald process.[9]

Britain depended heavily on imports of Chilean nitrate and its reaction with sulfuric acid throughout the war to manufacture nitric acid. The British, however, were well aware of the Haber-Bosch process. In 1917, a decision was made to construct a chemical plant at Billingham, England, for synthesising ammonia using the process, although the plant did not become operational until 1923.[10]

In 1914, the United States relied solely on Chile saltpetre and sulfuric acid for the manufacture of nitric acid. With the spectacular growth in the production of explosives during the war, however, supplies of sulfuric acid in the country began to dry up. Furthermore, to avoid reliance on imported supplies of sodium nitrate, the United States began to search for other ways of manufacturing nitric acid. One possibility was the Birkeland-Eyde process developed by Norwegians Kristian Birkeland (1866–1940) and Sam Eyde (1867–1917) in 1903. The method employed an electric arc to produce nitric oxide from nitrogen and oxygen. The oxide was then dissolved in water to form nitric acid. The method consumed a lot of power and was therefore relatively expensive, but was used to a limited extent by France and the neutral countries Norway and Sweden during the war.[11]

Another method of producing the acid relied on a supply of ammonia obtained by the reaction of calcium cyanamide, a nitrogen-containing compound, with water. The ammonia produced by the reaction could then be oxidised to nitric acid by the Ostwald process. The process also relied on electrical power, although much less so than the Birkeland-Eyde process.

Calcium cyanamide was manufactured by heating calcium carbide, a chemical compound made up of just calcium, another metal, and carbon, in nitrogen. It is produced by heating two relatively inexpensive materials, calcium oxide and coke, in an electric furnace. Calcium oxide, also known as lime, is obtained by heating limestone or other rocks containing calcium carbonate. The first plant for manufacturing nitric acid from ammonia using the cyanamide process was established in New Jersey, United States in 1916.[12]

By 1917, the United States was manufacturing nitric acid not only from Chilean nitrate, but also from ammonia obtained as a by-product from coke ovens, and, to a limited extent, from the ammonia generated from cyanamide. All three methods relied on the availability of platinum, either for the manufacture of sulfuric acid by the contact process, or for the oxidation of ammonia by the Ostwald process.

Until 1915, the United States imported most of its platinum from England and France, with about 93 per cent of this platinum originating from mineral deposits in the Ural mountains of Russia.[13] Typically, mining, milling and smelting one ton of the mineral yielded just a fraction of an ounce of crude platinum. The metal was then exported to England, France, Germany and other countries in Europe. Finally, companies such as Johnson Matthey in England refined the metal.

Platinum, like gold, silver and some other metals are known not only as precious metals but also as 'noble' metals. Unlike common metals and alloys that corrode in the presence of air and moisture, noble metals are highly resistant to corrosion. Corrosion is essentially the chemical degradation of a metal or alloy by moisture in the air or some other external agent in the environment in which the metal or alloy is used. The atmospheric corrosion of iron and steel is known as 'rusting' and results in the formation of a mixture of iron oxides commonly known as rust. The atmospheric corrosion of copper or copper-containing alloys like brass or bronze results in the formation of a green tarnish or patina known as verdigris, which consists of a mixture of copper carbonate, copper chloride and other copper compounds.

Because of its desirable chemical and physical properties, platinum was used not just for producing sulfuric and nitric acids but also for a variety of other civilian and military applications during the First World War. In chemical laboratories throughout the world, crucibles and wires used for chemical analysis were made of the metal; it was employed in dentistry to make pins for artificial teeth and for other purposes; the lead-in wires that carried the electric current into early electric lamps were made of platinum; it was widely used for telegraph and telephone electrical contacts; and in the ignition systems of the some of the earliest automobile internal combustion engines.[14] One of the earliest forms of photography produced platinum prints, also known as 'platinotypes'. The photographic process involved making paper sensitive to light by coating it with a mixture of two chemicals, one of which contained platinum. Finally, the metal was extensively used to make jewellery. All of these applications created an immense demand for the metal. In September 1915, England and France placed an embargo on the export of platinum in order to conserve supplies for the manufacture of the acids needed to make explosives.

The embargo starved the United States of its supply of platinum and by 1917, the year it entered the war, the problem was so acute that several national organisations launched initiatives to conserve the precious metal for the production of munitions. Delegates to a convention of the Daughters of the

American Revolution (DAR) pledged, in a formal vote, 'to refuse to purchase or accept as gifts for the duration of the war, jewellery and other articles made in whole or part of platinum, so that all available supplies of this precious metal shall be available for employment where they can do the greatest good in the service of our country'.[15] The DAR national society is a non-political volunteer women's service organization dedicated, among other things, to promoting patriotism and preserving American history.

The ACS inaugurated a 'movement for the conservation of platinum' at a meeting in Kansas City. The society also made the following appeal: 'Lend a strong hand to this important movement, wives and daughters of our chemists, in conversations with your friends, through club addresses, through your publications, and by public appeals wherever gatherings are held.' The initiative also involved searching for substitutes for platinum that could be used to make laboratory utensils such as crucibles.[16] The National Academy of Sciences, which like DAR and ACS is based in Washington DC, made a similar appeal.

The societies encouraged jewellers to 'overcome the abuse of platinum' by avoiding the use of the metal in bulky and heavy pieces of jewelry.[17] Jewellers were discouraged from using platinum 'in all non-essential parts of jewellery, such as scarfpin stems, pin tongues, joints, catches, swivels, spring rings, ear backs etc., where gold would satisfactorily serve.' However, jewellers 'fought strenuously to keep a supply available for their trade,' according to one report.[18]

By the beginning of 1918, the Women's National League for the Conservation of Platinum had been established in the United States.[19] Soon after, the United States government issued an order to commandeer all crude and unworked platinum in the hands of importers, jobbers and wholesalers for the manufacture of munitions.[20]

The Second Industrial Revolution

In many ways, the First World War rode on the back of the technological revolution that occurred in the second half of the nineteenth century. The revolution, sometimes referred to as the Second Industrial Revolution, was part of the Industrial Revolution that started in Britain about the middle of the eighteenth century and in the following decades in other European countries, as well as in the United States.

The first part of the Industrial Revolution was characterised by the growth of steam power, mechanisation, manufacturing industry and the mass production of goods, notably textiles, transforming countries from rural to urban economies.

During the second part of the revolution, the numerous advances in science, technology and engineering that occurred during the nineteenth century led to the mass production of the arms and munitions on which the First World War depended, and to the rapid developments in railways and other transport systems for carrying soldiers and their equipment to the front. These developments were accompanied by advances in the use of electricity and communication systems such as the telephone and wireless telegraphy. The surge in the development and manufacture of synthetic organic chemicals, especially dyes and pharmaceuticals, photographic chemicals and other types of chemicals, also occurred during the second industrial revolution.

Both parts of the Industrial Revolution were underpinned by two chemical elements: iron, a metal, and carbon, a non-metal. The use of carbon, in the form of charcoal, to extract iron from the minerals in ores dates back some 3,000 years to the Iron Age.

Minerals are naturally occurring materials that make up the Earth. They are crystalline substances with distinct chemical structures and characteristic chemical compositions, and many, but not all, contain metals. Ores, on the other hand, are rocks in the Earth's crust that contain one or more minerals from which metals can be extracted. Coal is a type of rock, but it is not an ore as metals are not extracted from it, although it does contain small amounts of minerals.

Large deposits of ores containing the mineral hematite, which has the chemical name ferric oxide or iron(III) oxide, are found in many parts of the world. The iron is extracted from the ore by smelting, a chemical process in which the carbon combines with and removes the oxygen from the mineral to leave behind the metal. A type of furnace known as a 'bloomery' was originally used to smelt iron ore and produce iron. The process yielded a crude, impure form of the metal known as a 'bloom', which could then be forged into wrought iron. This type of iron is a tough, almost pure, form of the metal and contains very small amounts of carbon and other impurities.

From the late fifteenth century onwards, blast furnaces began to replace bloomeries, especially in Europe and, by the time of the Industrial Revolution, they were being widely used for the production of iron. The carbon for the blast furnaces was usually supplied in the form of coke, a material obtained by heating coal in an oven to remove coal tar. Alternatively, anthracite – a type of coal with a high carbon content – was used to supply the carbon.

Blast furnaces are still used today to smelt iron ore. In a typical process, iron ore, coke and limestone are added to the top of the furnace while air is blown into the bottom. The limestone, which is a form of calcium carbonate, reacts with impurities, such as sand, in the ore to produce molten slag. The slag floats to the top of the molten iron. The molten iron beneath the slag is drawn off at the bottom of the furnace and run into moulds of sand. This type of iron, known as pig iron, is then re-melted, often along with scrap iron, to make cast iron, a crude, impure form of iron containing 2–4 per cent carbon, as well as smaller amounts of silicon and other elements.

During the early decades of the Industrial Revolution, wrought iron was produced from cast iron by a process known as 'puddling'. The process involved heating and melting the cast iron in a puddling furnace and stirring the molten metal with a 'puddling bar'. Carbon-, silicon- and manganese-containing impurities in the cast iron were converted into oxides such as carbon dioxide and silicates and removed from the furnace either as exhaust gases or as slag. The purified iron, which by now had become spongy, was separated from the slag and removed from the furnace as 'puddle balls'. Whereas cast iron is brittle and liable to fracture, wrought iron is malleable which means that it can be hammered, stretched or twisted into various shapes and objects. For example, two or more pieces of the metal can be welded together by hammering at high temperatures.

The puddling process was used to a limited extent to produce mild steel, a type of carbon steel that contains less than 0.3 per cent carbon. Carbon steels are alloys of iron and up to 2 per cent carbon, and like wrought iron, which also has a very low carbon content, it is also very malleable. Wrought iron, however, contains minute amounts of slag, known as 'stringers', which gives it its characteristic fibrous structure.

Other processes were also used to produce steel. For example, crucible steel, a high-quality and expensive steel, was typically manufactured by charging a clay crucible in a coke-fired furnace with inferior steel and a chemical agent known as a 'flux'; the agent combined with the slag and other impurities in the steel to form a liquid which floated to the surface of the molten metal. After heating for several hours, the crucible was removed from the furnace, the impurities skimmed off and the molten steel cast into moulds to form ingots. Crucible steel was used to make cutlery and a variety of luxury products.

Before the mid-nineteenth century, steel was expensive and so relatively little was produced. Cast iron and wrought iron, however, were produced on a large scale for a multiplicity of applications. They were used, for example, to make railway lines, girders for bridges, boilers and ships' plates. The earliest steam locomotives, built in the first half of the nineteenth century, ran on cast iron rails and later on wrought iron rails.

The large-scale manufacture of ferrous metals, that is iron and steel, relied on a steady and plentiful supply of iron ore and coal. Coal was needed not just to supply the carbon for steel but also to fuel the furnaces. Countries such as Britain, Germany and particularly the United States were richly endowed with both materials and were therefore in ideal positions to exploit developments in ferrous metals technology.

A major breakthrough occurred in 1855 when Henry Bessemer (1813–1898), an English metallurgist and inventor who had been trying to develop methods for making stronger gun barrels, patented an economic process for the mass production of steel. Bessemer showed that molten pig iron could be converted relatively quickly into low-cost, high-quality steel using a large tilting vessel known as a 'Bessemer converter'. The converter was first tilted and molten pig iron poured into an opening at the top. The vessel was then uprighted and air blown in from the bottom through the molten iron to oxidise the impurities. Finally, the converter was tilted again to discharge the molten steel. The amount of carbon in the steel could be adjusted as required by varying the length of time the air was blown into the vessel. The steel was shaped by machining, by forging – with a hammer for example – or by casting the molten metal into a mould.

The Bessemer process lowered the cost of steel from around £50–£60 per ton to £6–£7 per ton.[21] Other low-cost methods for making high-quality steels soon followed, the most important of which was the open-hearth process developed by German-born British engineer William Siemens (1823–1883) and French engineer Pierre-Emile Martin (1824–1915) in the 1860s. The Siemens-Martin process, as it is also known, employed a furnace with a shallow hearth or basin. The hearth was filled with similar amounts of molten pig iron from a blast furnace and scrap iron and steel, together with small amounts of iron ore and limestone. The molten steel formed as a shallow pool in the hearth. It had several advantages over the Bessemer process in that it produced stronger steel and could also refine iron containing large amounts of phosphorus impurities, although, on the downside, it was slower than the Bessemer process.

Another form of carbon, known as graphite, also played a key role in steel making in the years leading up to the start of the First World War. The electric arc furnace, invented by French metallurgist Paul Héroult (1863–1914), first installed in California, United States in 1907 and still used today, employs graphite electrodes to convert scrap iron into steel. The furnace is charged with scrap iron and a material such as limestone that converts impurities into slag. The electrodes are lowered into the scrap and an arc of electric current is passed through the material. The process creates very high temperatures that melt the scrap and convert the impurities to oxides. The molten steel and molten slag are then tapped off separately at the bottom of the furnace.

With these advances in technology, steel making rapidly became big business. Following the introduction of the Bessemer and open-hearth processes, low-cost, high-quality steel began to replace cast and wrought iron in the manufacture of armaments and railway tracks, and as a construction material for ships, bridges, high-pressure boilers, steam engines and turbines for power generation throughout Europe and the United States.

In 1859, Bessemer himself established a steel works at Sheffield, England, while other steel producers soon followed. For example, the Midvale Steel Works began operating in Philadelphia, Pennsylvania, in 1867. William Beardmore, a steel-making company that manufactured armaments, locomotives, and ships was founded in Scotland in 1887, and the United States Steel Corporation was established in Pittsburgh, Pennsylvania in 1901.

In 1915, Geo W. Sargent, vice-president of the Crucible Steel Company of America, produced statistics in the *Journal of Industrial and Engineering Chemistry* showing that total steel production in the United States in 1898 was just over 8.9 million tons, of which roughly 6.6 millions tons were made by the Bessemer process, 2.2 million tons by the open-hearth process, and the remainder by the electric arc, crucible and other processes.[22] By 1913, the year before the start of the First World War, the total had risen more than three-fold to over 31.3 million tons, of which approximately 9.5 million tons were made by the Bessemer process, 21.6 million tons by the open-hearth process and the remainder by the other processes. These statistics show that the Bessemer process was the dominant method of steel production in the country at the end of the nineteenth century, whereas fifteen years later the open-hearth method accounted for about two thirds of all production. The electric arc, crucible and other methods remained niche processes for producing high-quality steels throughout this period.

In the same issue of the journal, Allerton S. Cushman, director of the Institute of Industrial Research, Washington DC, comments that 'unless the chemists of Germany had worked out the synthesis of nitrogen compounds from atmospheric nitrogen as well as the special steels suitable for construction of the necessary apparatus, the present great world war would have to be brought to an early close'.[23]

Ammonia was one of the nitrogen compounds to which Cushman refers. The Haber-Bosch process that was used to manufacture ammonia employed iron not only as a catalyst but also to construct the equipment for converting the nitrogen and hydrogen into ammonia. The converter had to withstand the high temperatures and pressures necessary for the reaction to occur. However, the iron used to make the original pilot version of the converter contained too much carbon, so the converter became brittle and several explosions occurred.[24] Bosch solved the problem by constructing a converter consisting of a tube made of soft steel with a low percentage of carbon, which was then placed inside a tube of ordinary steel.

The course and duration of the First World War relied not just on the Haber-Bosch process and the iron and steel needed for its operation, but also on the manufacture of numerous other iron and steel products. They included guns, shell casings, barbed wire, armour plate for battleships and tanks, protective helmets, corrugated iron, steam locomotives, railway lines, gas cylinders and shipping containers.

The use of steel for the manufacture of guns can be traced back to the years before Bessemer introduced his process for manufacturing steel. During the middle of the nineteenth century, German arms manufacturer Alfred Krupp (1812–1887) experimented with various types of steel for musket and cannon barrels. In 1847, he manufactured the first steel gun – a cannon of cast steel. Four years later, he exhibited a solid, cast steel ingot weighing two tons at the Great Exhibition in London and subsequently constructed the first Bessemer steel plant.

By the start of the First World War, steel manufacturers were using the Bessemer, open-hearth and others processes to make not only carbon steels, such as mild steel, but also alloy steels containing metals such as chromium, manganese and nickel. These non-ferrous metals were added to improve the properties of steel. For example, whereas carbon steels rust in moist air, alloy steels, such as stainless steels, do not. Stainless steels contain at least 10 per cent chromium and usually have small amounts of other metals, such as nickel. Chromium hardens the steel and increases its resistance to corrosion by forming a thin protective coating of chromium oxide on the surface of the steel. Manganese was also widely used to harden steels and reduce their brittleness, while the addition of nickel to a carbon steel also increases its toughness.

The use of alloy steels to make armaments dates back to the late nineteenth century when Krupp began to manufacture howitzers with barrels fabricated with a nickel alloy steel that was not only strong but also light. The firm soon became the world leader in the development and manufacture of alloy steels for a variety of applications. Krupp cemented armour, for example, was made of a hard alloy steel containing nickel, chromium and manganese. The armour was widely used by the German and British navies to protect their battleships.

A variety of steels were also used as armour for the American, British, French, German and Russian tanks that entered combat from 1916 onwards. One of the earliest tanks, the prototype British Mark I, which was designed in 1915 and completed in January 1916, employed nickel alloy steel plate as protective armour.[25]

Steel armour was used in the First World War not just on warships and tanks, but also for personal protection, most notably in the form of helmets. French infantry troops in the trenches were the first to be provided with helmets made of inexpensive mild steel. The helmets, issued in 1915, were designed to prevent head wounds caused by shrapnel from shell bursts. The steel of the helmet was enamelled a dull blue. Enamel, a form of glass and coating, not only hardened the outer surface of the steel helmet but also kept it smooth, scratch-resistant and easy to clean. The helmet was called the 'Adrian' helmet after the French engineer and army officer, August-Louis Adrian (1859–1933), who designed it.

Later in 1915, the British introduced the 'Brodie' helmet, made of mild steel and named after its British designer, John L. Brodie. It was subsequently replaced by a helmet made of Hadfield's steel, a harder manganese-containing steel that is highly resistant to wear. The steel, also known as manganese steel, was discovered by British metallurgist Robert Hadfield (1858–1840) in 1882, and it was the first alloy steel to be produced from carbon steel. Four years later, Hadfield patented silicon steel, an alloy steel containing silicon, a type of chemical element known as a metalloid which has the characteristics of both a metal and a non-metal. Similarly, in 1916, the German Army introduced the *Stahlhelm*, a helmet made of an alloy steel containing both nickel and silicon, and even harder than manganese steel.

The use of helmets in the war dramatically reduced the incidence of head wounds, according to a *British Medical Journal* correspondent in Northern France. In an article on armour published in April 1916, the un-named correspondent suggests that, before the introduction of steel helmets, wounds of the head, neck and face accounted for about 25 per cent of all wounds incurred by British troops who survived long enough to reach the casualty clearing stations.[26] The correspondent reports that, following the introduction of helmets, of 1,000 casualties arriving at a particular casualty clearing station following a battle at the beginning of March 1916, 'the number of penetrating gunshot wounds of the head was equal to less than ½ per cent, while the total number of fractures of the skull was well under 1 per cent of all injuries'.

Other alloy steels were also of critical importance during the war. For example, steels containing the metallic element tungsten were used to make the cutting tools and drill bits needed in engineering works and armament factories. Steel containing another metallic element, molybdenum, was used as a lining for large guns.

France, Germany and Britain could not have mobilised troops and rapidly brought them to the Western Front without the steels needed to manufacture railway tracks and the bodies, wheels and other components of steam locomotives. The trains ran on railway lines made from the exceptionally durable manganese steel developed by Hadfield.

Countless other iron and steel products were exploited in the war. The chemical warfare on the Western Front, for example, relied extensively on the manufacture of iron and steel, and not just for the fabrication of the guns and shell cases that contained the chemical warfare agents. The cylinders used to store and transport the liquefied chlorine, mustard gas, phosgene and other poisonous gases were made of cast iron or steel. Steel barrels and drums were also used to ship inflammable and volatile organic liquids such as gasoline and liquefied petroleum gas.[27]

Corrugated iron is another example. The material was widely used to shore up the sides of trenches and as a roofing material. It was produced by dipping sheets of mild steel in a bath of molten zinc – another metallic element. The process, known as galvanisation, prevents the steel rusting.

The metals added to carbon steel to make alloy steels and to galvanise iron are all extracted from ores by chemical processes similar to those used to smelt iron ore. Nickel, for example, is obtained by roasting ores containing a mineral such as pentlandite with coke. Pentlandite, named after the Irish scientist Joseph Pentland (1797–1873) who first recorded it, contains the chemical elements iron, nickel and sulfur. The iron is removed as slag, leaving nickel sulfide which is then roasted in air. The oxygen in the air converts the sulfide into nickel oxide. Finally, the carbon in the coke combines with the oxygen in the oxide to form carbon dioxide and leave nickel. This part of the smelting process is known as reduction. Chromium, manganese and tungsten are extracted from metal oxide ores in similar fashion to the extraction of iron from iron oxide ores.

The First World War severely disrupted the supply of ores around the world and the production of the metals required by the belligerent nations to sustain their war efforts. Zinc, or spelter as it is also known, was a particular example. It was used not only to galvanise iron, but also to manufacture alloys such as brass and for other applications. The metal was extracted by smelting ores containing zinc sulfide minerals. At the start of the war, the annual production of zinc was around 1 million tons.[28] Germany and Belgium accounted for about two thirds of this amount and the United States for the rest. After Germany occupied Belgium in 1914, the Allies became dependent on the United States for its supplies of zinc. By the end war, the United States had doubled its annual production of the metal.

Metal-containing minerals in ores were a source of not just the metals needed to make iron and steel products, but also a source of some the chemicals that underpinned the operation of the First World War. For example, iron pyrite, one of the most abundant and widespread of all sulfide minerals, was, as we saw above, one of the sources of the sulfur needed for the manufacture of sulfuric acid. Furthermore, the slag that contained the impurities from ore smelting could also be put to good use. A report in 1914, before the outbreak of war, observes that the use of iron-furnace slag was well developed in Germany: 'In solid form it serves for highway and railway construction and as a body material for concrete'.[29] Granulated slag could also serve to make building blocks, bricks and tiles, and for the manufacture of cement. Finally, slag was employed as a raw material to make glass, artificial marble and artificial pumice stone.

In 1995, American scientist Kevin K. Olsen argued that 'without ferrous metals technology, much of the modern world simply would not exist.'[30] The same argument could well be extended to the First World War. The war, with its industrial-scale carnage and destruction, could not have occurred without the large-scale manufacture of the iron and steel needed for the construction of ships, railways, guns, ammunition, tanks and other articles of war.

Non-ferrous metals were also essential

Non-ferrous metallic chemical elements were essential not just for the manufacture of the alloy steels that underpinned much of the military efforts in the First World War, they also played other key roles. A brief report on British metals and munitions, published in the United States in February 1916, emphasises the importance of these metals.[31] It comments that non-ferrous metals such as nickel, copper and zinc were of such 'enormous value in the production of munitions' that it was not surprising that the London-based Institute of Metals 'should be especially interested in the production of the numerous metal articles required for the prosecution of the war'.

A few months after this report was published, a German merchant submarine, the *Deutschland*, broke through the British naval blockade on her maiden voyage. The U-boat arrived in Baltimore in the United States, which had yet to enter the war, with a cargo of chemicals, dyes, pharmaceutical products and precious stones.[32] In early August 1916, the vessel set off for Bremerhaven, Germany, carrying hundreds of tons of crude rubber, nickel and tin.

Both metals were important for the war effort, most notably for the manufacture of ferrous and non-ferrous alloys. As previously mentioned, nickel was a component of the alloy steels developed by Krupp. Tin was also used to make a range of non-ferrous alloys. For example, it was mixed with lead, a soft malleable metallic element, to make lead solder, also known as soft solder as it is relatively easy to melt. It has a lower melting point than hard solder, an alloy that typically contains copper and zinc or silver. Solder, in one form or another, has been used for thousands of years to join pieces of metal together.

Tin was also alloyed with copper to make bronze, which has been employed to make tools and weapons for over 5,000 years since the so-called Bronze Age. The term 'bronze' is also applied to copper alloys that do not necessarily contain tin, but may contain other metals such as manganese or aluminium. Bronze had been used for centuries to make cannon barrels, but by the time of the First World War the alloy had been replaced by steel. Even so, special types of bronze were still used in the war to make components for armaments and for other purposes. For example, phosphor bronze, which is a tough corrosion-resistant alloy containing not only copper and tin but also a small amount of the non-metallic element phosphorus, was used to make springs and bolts. During the First World War, phosphor-bronze springs were employed in certain types of percussion fuses that were screwed into the bases of armour-piercing and other types of shells.[33] The spring prevented the detonation of the fuse before impact.

Manganese bronze, an alloy of four metallic elements – copper, zinc and small amounts of manganese and iron – was also widely used in the war. Both manganese bronze and phosphor bronze do not corrode in seawater and were therefore used to make ships' propellers. The largest passenger steamship in the world, the *Titanic*, which set sail on its fateful maiden voyage from Southampton on 10 April 1912, was propelled by three large manganese bronze propellers.[34] The *Lusitania*, sunk off the south coast of Ireland by a German U-boat on 7 May 1915, was propelled by four 15-ton manganese bronze propellers.

Copper and zinc were not only components of manganese bronze, but also of another well-known alloy, brass. The alloy typically consists of 70 per cent copper and 30 per cent zinc, while some types also contain other metals such as lead or aluminium. The alloy is a relatively soft and malleable material that, when molten, can be readily cast into a shape by pouring the liquid into a mould. The alloy was used to make various small components of fuses and ignition tubes, and the cartridge cases for the rifles, pistols and other small arms used in the

First World War. Each brass cartridge case packaged the three key components of ammunition: a primer, a propellant and a projectile. Brass cartridge cases were also used to package the propellants for quick-fire artillery weapons.

Gunmetal is an alloy which, in terms of its chemical composition, is a cross between brass and bronze. It typically consists of 88 per cent copper, 10 per cent tin and 2 per cent zinc. The alloy, as its name implies, was once used for making guns before the advent of low-cost steels during the Second Industrial Revolution. Although gunmetal is only about half as strong as mild steel, it was employed in the war for a variety of purposes; for example, to make the threaded sleeves which the fuses were screwed into in the noses of some types of high-explosive shells.[35]

The copper used to produce bronze, brass and gunmetal was obtained by smelting copper-containing ores, the most abundant of which is an ore containing a mineral known as chalcopyrite, also known as copper iron sulfide. The copper is extracted from the mineral in a process that is similar to that used to extract nickel from sulfide ores. Lead was also obtained from ores containing sulfide minerals of which galena, or lead sulfide, is the most widespread.

Some brasses, bronzes and gunmetals also contained lead, the presence of which enabled the alloys to be more easily machined. One such alloy was red brass which contained 85 per cent copper, 10 per cent zinc, 5 per cent tin and 5 per cent lead.[36]

The tin needed to make bronze and gunmetal was usually extracted from an oxide mineral known variously as cassiterite, tinstone or tin dioxide. The manganese of manganese bronze was typically extracted in the same way from ores containing manganese oxide minerals, of which the most important was pyrolusite, also known as manganese dioxide.

The use of non-ferrous metals was not confined to the manufacture of alloys, however, as they also had many other uses. Mercury, also known as quicksilver, is the only metallic element that is liquid at room temperature and pressure, and provides a unique example. As we saw in the previous chapter, a compound containing the metal, namely mercury fulminate, is a primary explosive that was used to fire propellant charges and detonate bursting charges. Mercury fulminate required three chemicals for its manufacture: mercury, nitric acid and ethanol.

The use of mercury fulminate to manufacture munitions was historically a perilous activity. In one horrific incident, chemist Henry Hennell was 'blown to pieces' in the Apothecaries' Hall, London, on 4 June 1842 as he mixed 6lb of the compound.[37] Hennell, a Fellow of the Royal Society, was mixing portions of the powder in a china bowl to fulfil 'a large military order'.[38] He was making a bomb to be used in the first Anglo-Afghan war that was fought from 1839 to 1842. At the inquest on 6 June 1842, it was stated that the upper part of his body was shattered and 'the fragments cast to a distance'. An arm was found on the roof of the hall and a finger picked up in a street more than a 100yd away.

Mercury is one of the few metals that occurs naturally in its native state and not chemically combined with other elements; other such metals are gold, silver, platinum and copper. Mercury is more commonly found in mercury-containing minerals such as mercury sulphide, which is found in an ore called cinnabar. The metal is separated from the sulfur by roasting the crushed ore in a furnace, which evaporates the metal before it is condensed back into a liquid.

Metals that took a starring role

The manufacture of star shells used by both sides to illuminate each other's positions in the frontline provides a prime example of how the military hardware and ammunition used in the First World War depended on the production of a wide variety of both ferrous and non-ferrous metals, and metal-containing chemical compounds.

The star shells fired by the BL 6in howitzers were typical: the shell had a steel body and steel head that was attached to the body by six brass screws and six steel twisting pins; a wooden block inside the steel head had a gunmetal socket to take the fuse; a wrought iron washer separated the block from the body; a two-part wrought iron central tube, containing a fine-grain gunpowder primer in a fabric bag, was separated from the bursting charge by a wrought iron disc; the bag was kept in position by a piece of copper wire; the tube itself consisted of two parts, each of which was surrounded by a tier of stars; the two tiers were separated by a perforated iron disc and supported by corrugated steel supports locked into place by wrought iron pins.[39]

The shell could not be fired without the assistance of another metal – potassium. It was present in the potassium nitrate, one of the components of the gunpowder bursting charge that ejected the stars from the shell. Although potassium is a highly reactive metal, it plays a passive role in the explosive. The active component of the compound is the nitrate, which oxidises the carbon and sulfur in the gunpowder, resulting in the release of energy and the production of gases and smoke. The potassium itself ends up in the smoke as a component of chemical compounds such as potassium sulfide, potassium carbonate and potassium sulfate. These compounds are soluble in water, so they eventually get washed away.

Most importantly, however, the stars used with these shells and also for signal rockets relied on metals for the colours of their flares. Generations of chemistry students will remember carrying out the flame test, a test that works on a similar principle to that of the star shell. The aim of the test is to identify the metal components of salts. The tip of a clean metal wire secured in a glass rod is dipped into a sample of the salt and then placed in a Bunsen burner flame.[40] These salts produce different colours in the Bunsen flame: sodium salts such as sodium chloride (common table salt) are golden yellow; potassium salts such as potassium chloride are lilac; and calcium salts such as calcium chloride burn a brick red colour.

Metal salts are used as colour agents in the pyrotechnic mixtures in fireworks.[41] For example, a range of sodium salts are used to produce yellow effects. Similarly, the stars in the star shells and signal rockets used in the First World War contained colour agents. A typical star composition contained potassium chlorate, then known as 'chlorate of potash', as an explosive material, shellac as a fuel that burns at low temperature, and a colour agent. The salts copper carbonate, strontium carbonate and barium chlorate were used as colour agents to produce blue, red and green flares respectively, while star compositions containing magnesium powder burnt with a bright white light.[42]

As we saw in chapter two, various metallic components were required for the fabrication of fuses. The 1915 British War Office *Treatise on Ammunition* provided details of many of the fuses used in the war, including a time fuse for star shells.[43] The body and time ring of the fuse were made of aluminium, and the fuse also contained steel, brass and phosphor bronze components. In addition, all British fuses were packed into tin cylinders to protect them from the damp and prevent deterioration. Lids were fixed onto the cylinders using solder, which was made from lead and tin.

So, if a chemist were to analyse the metallic content of a fused green star shell, how many metallic chemical elements would he or she find? Iron would be present in the wrought iron and steel parts of the shell and aluminium in the fuze body; copper, zinc and tin would be found in the copper, brass, gunmetal, bronze and corrugated iron components; and finally, potassium and barium would be identified in the chemical compounds in the bursting charge and the star composition respectively. That makes a total of seven metallic elements present in one fused shell.

And let's not forget bullets

In December 1914, several German surgeons waxed 'eloquent on the virtues of the French bullet' according to a report in the *British Medical Journal*.[44] The French rifle bullet was made of solid copper, although an earlier report in the same journal states that it was made of bronze containing 90 per cent copper.[45] The surgeons claim that the solid copper bullet inflicted 'as small and neat a wound as is possible under the circumstances', causing little damage, even when it penetrated bones or vital organs.

The German 'S' rifle bullet on the other hand produced 'unpleasant wounds' observed John Bland-Sutton (1855–1936), surgeon to the Middlesex Hospital, London, in a lecture on the diagnosis of bullet wounds that he delivered at the hospital in 1914.[46] The bullet, also known as a 'spitzer' after the German term *Spitzgeschoss*, meaning 'pointed bullet', was introduced in 1905.[47] The 'S' bullet consisted of a solid core of lead enclosed by a case made of an iron-nickel alloy.[48] The 'troublesome and often painful wounds,' explained Bland-Sutton, were caused by the bullet case pealing off into twisted and contorted shapes when the bullet hit and glanced off a hard object such as a stone. Furthermore, he added, the lead core of a glancing bullet broke up into tiny fragments that 'have destroyed many eyes'.

The corresponding British pointed rifle bullet, the Mark VII, used in the First World War, consisted of a copper-nickel alloy sheath which enclosed a core of two parts: the core in the front pointed section of the bullet was made of pure aluminium or an alloy of 90 per cent aluminium and 10 per cent zinc; the lower section of the core consisted of an alloy of 98 per cent lead and 2 per cent of the metallic element antimony.

The modern pointed nickel-sheathed rifle bullet 'was probably the most humane projectile yet devised', remarks eminent British physiologist and surgeon Victor Horsley (1857–1916) in a memorandum to the British War Office in 1914.[49] Horsley, who was to die of heat exhaustion two years later while on field surgery duty in Mesopotamia, notes that the sheath prevented the bullet deforming and breaking up into fragments except, as Bland-Sutton also observed, after a ricochet. Yet, according to a report published at the start of the war, the pointed bullet was not designed with a sheath primarily for humanitarian reasons. Rather it was designed to prevent the lead, a soft metal, from being torn from the bullet and fouling the grooves cut into the bore of the rifle.[45]

Until the beginning of the nineteenth century, bullets were generally made of lead, a metallic element that is not only soft but also heavier than many other metals. Lead also has a relatively low melting point, so the surface of a lead bullet therefore tended to melt when fired. The problem was overcome either by alloying the metal with another metal, such as tin, or by encasing the lead core with a hard metal sheath. Fully-sheathed bullets are known as 'full metal jacket' bullets and are designed not to expand when they hit their target.

An expanding bullet, known as a 'hollow' point bullet, on the other hand, enables the soft lead core to mushroom out of the case on impact and inflict more severe injuries. This mushrooming effect is achieved by hollowing out a pit in the tip of the bullet. Alternatively, the leaden core is exposed as a flat surface in the nose of the bullet. In this case, the bullet is known as a 'soft point' or 'soft nosed' bullet. Soft nosed and hollow point expanding bullets were developed in the 1890s at the British arsenal at Dum Dum, a city in West Bengal, India. For this reason, the bullets are sometimes called dumdum bullets.

Horsley suggests that the British and German pointed bullets conformed to the provisions of the Hague Conventions of 1899 and 1907 which banned the use in warfare of bullets that expand or flatten inside the human body. He remarks, however, that there were suggestions that both German and British soldiers had used the prohibited expanding bullets in the early months of the war.[49] That possibility led to some debate. Bland-Sutton, for instance, observed in his lecture that some of the injuries caused by 'S' pointed bullets that had been deformed by hitting hard objects were 'erroneously attributed to expanding bullets'.[46]

Rifle bullet wounds in soft tissue were similar in nature to shrapnel bullet wounds, observes Royal Navy surgeon A. Gascoigne Wildey in a report sent to the *British Medical Journal*.[50] His report describes the treatment of over 950 wounded Belgian soldiers received at the Royal Naval Hospital, Haslar, Gosport, England, in the middle of October 1914. Interestingly, shrapnel bullet wounds occurred in about the same proportion as rifle bullet wounds.

Shrapnel bullets were made as small as possible so that the maximum number could be packed into a shell. Like other bullets, they had to be sufficiently heavy to achieve the necessary velocity to be effective at killing or injuring enemy troops. Early shrapnel bullets were made of cast iron and known as 'sand shot'.[51] Pure lead, although a heavy metal, was too soft to be used for shrapnel bullets. Those used in the First World War typically consisted of a hard lead alloy consisting of one part antimony and seven parts lead. Antimony, like silicon, is a chemical element that has some metallic and some non-metallic properties, and is known as a metalloid. The element is extracted from ores containing sulfide minerals such as antimony sulfide, commonly known as stibnite.

A report in a December 1917 issue of the *British Medical Journal* notes that shrapnel bullets contained arsenic as well as lead and antimony; all three elements are highly toxic.[52] Arsenic, another metalloid element, was used to strengthen the lead-antimony alloy. The report outlines efforts to determine the solubility of these elements in body fluids by placing pieces of shrapnel bullets in human pus, blood serum of sheep, olive oil and other fluids. The experiments found significant amounts of lead but only traces of arsenic and antimony in the fluids: 'So far as the retention of shrapnel in the body is concerned, no damage is to be apprehended from arsenic and antimony … The one danger is lead'.

Aluminium – renowned for its lightness

In many ways, the metallic element aluminium is completely different from lead. Its melting point of 660°C is over twice that of lead (327°C). The difference in their densities is even more pronounced, with lead being more than four times heavier than aluminium. Both elements, however, are soft and malleable, and were used for a variety of applications in the First World War.

As previously mentioned, aluminium was used in parts of the cores of Mark VII pointed bullets, the bodies and time rings of fuses used for star shells, and the powdered form was a component of the explosive ammonal and the incendiary material thermite. However, one of its most important uses was in the construction of the airships used in the war. Airships are motorised aircraft that can be propelled and steered through the air, while their cousins, observation and barrage balloons, are not classified as airships as they rely on wind for their movement.

The airship was first developed by the French in the late eighteenth century and was called a *ballon dirigible*, or steerable balloon. There are two basic types of dirigible: the rigid, which has a rigid framework, and the non-rigid which does not. A third type, the semi-rigid airship, is a cross between the two.

The rigid airships employed in the First World War had plywood or metallic cage-like frameworks covered with fabric. The airships were propelled by engines with propellers, the engines being attached to the undersides of the structures. The framework enclosed a series of balloons, called gas cells, filled with hydrogen, the lightest of all gases. Hydrogen, however, is highly flammable, so nowadays helium, a non-flammable gas, is used even though it has almost twice the density of hydrogen.

The zeppelin, the most famous of all the rigid airships used in the war, was invented and developed by Germany Army officer Count Ferdinand von Zeppelin (1838–1917) between 1897 and 1900. The first zeppelin, known as *LZ1*, made its brief maiden flight from Lake Constance in Germany on 2 July 1900, carrying five passengers. The airship had an aluminium frame and contained sixteen cells of hydrogen gas.

The German Army and Navy employed well over a hundred zeppelins for reconnaissance and bombing missions during the war, and the majority of these were constructed during the war. Whereas *LZ1* had a pure aluminium framework, subsequent zeppelins were built with aluminium alloy frameworks. One of the most important of these alloys consisted of aluminium and small amounts of copper, manganese and magnesium. The alloy, known as duralumin, was patented by German metallurgist Alfred Wilm (1869–1937) in 1906.

The first zeppelin bombing mission was carried out in August 1914 when *LZ21* dropped artillery shells on Liege in Belgium. Zeppelins subsequently bombed various towns and villages on the east coast of England in the early months of 1915, while the first attack on London occurred on 31 May that year, killing seven and injuring thirty-five. In that raid and three other raids on London during 1915, a total of 278 bombs were dropped on the city, of which 198 were incendiary bombs.[53] The incendiaries contained white phosphorus and thermite – a mixture of aluminium powder and iron oxide that burns ferociously when ignited.[54]

By mid-1916, British planes were beginning to destroy zeppelins with incendiary bullets such as the 'Pomeroy' bullet and the 'Buckingham' bullet. These bullets were fired from machine-guns either over the wings or to the rear of the aircraft, exploding on impact, igniting the hydrogen and causing the airship to crash in flames.[55]

The Pomeroy bullet, named after its inventor New Zealand-born engineer John Pomeroy (1873–1950), had a hollow nose containing a charge of dynamite. The Buckingham bullet, patented in early in 1915 by British chemist James Buckingham, was filled with a phosphorus-containing composition.

Versions used later in the war consisted of an envelope, which contained the phosphorus composition, made of a copper-nickel alloy into which two lead plugs were inserted.[56] When the bullet was fired from a gun, the phosphorus melted and was ejected from the bullet, igniting spontaneously when it came into contact with air.

The British built only eight rigid airships that saw service during the war. They were mostly modelled on German zeppelin designs but proved to be of limited use.[57] The first, known as *No. 9*, entered service in April 1917 and, like the zeppelins, it had a duralumin framework. The airship was employed for experimental and patrol work, and for training crews of future rigid airships.

In the early years of the war, Britain relied on non-rigid airships, fondly known as 'blimps', rather than on rigid airships to serve with the Royal Navy. Britain had begun the war with just seven blimps, but when German submarine warfare against merchant shipping began in February 1915, Britain rapidly developed non-rigid airships known as the Submarine Scout or SS class blimps to carry out anti-submarine patrols. A total of 158 of these airships were built.[58]

Although blimps did not have rigid frameworks, aluminium was used in their construction in various ways. For example, the envelopes that contained the gas were made of a rubber-proofed fabric coated with four layers of dope, including two layers of aluminium dope, and one layer of aluminium varnish. The coatings protected the envelopes from the adverse effects of weather. Aluminium was also used to make various parts of the blimp, including some of the gas valve components, air tubes, fins and fuel tanks.

Aluminium is the most abundant metal on Earth and the third most abundant element after oxygen and silicon, both of which occur in silicate minerals. Aluminium makes up about 8 per cent of the Earth's crust whereas oxygen and silicon make up approximately 47 per cent and 28 per cent respectively. The aluminium metal needed for the war effort was extracted from bauxite, the most widespread of all aluminium ores. Bauxite is a crude mixture of hydrated alumina minerals and impurities such as iron oxide and silicates. Alumina, also known as aluminium oxide, is a compound of aluminium and oxygen in the ratio two to three, while the hydrated forms of the compound also contain hydrogen.

The process for extracting aluminium from alumina, which is still used today, was developed during the Second Industrial Revolution. The alumina in the mineral is first separated from the impurities and purified in a process patented in 1888 by Austrian chemist Karl Bayer (1847–1904). The crude ore is first crushed and mixed with sodium hydroxide (also known as caustic soda) solution. The mixture is pumped into a 'digester' where it is heated to about 175°C under high pressure.[59] The iron oxide is then removed from the mixture as 'red mud' by allowing it to settle out, while the solution is filtered to remove other solid impurities and transferred to a tank where aluminium hydroxide separates out as a white crystalline solid. Next, the solid hydroxide is filtered from the solution and washed. Finally, the hydroxide is roasted in a kiln at about 1,000°C to drive off water as steam to form pure alumina.

The pure alumina is then melted and aluminium separated from the molten material by electrolysis. As the melting point of alumina is 2,000°C, electrolysis of pure molten alumina is neither practicable nor economic, as the energy costs would be prohibitive. The melting temperature is therefore lowered by mixing alumina with cryolite, a mineral that contains aluminium, sodium and fluorine. A mixture containing 5 per cent alumina and 95 per cent cryolite melts at 970°C.

The electrolysis is carried out in a large vessel containing carbon electrodes known as a Hall-Héroult cell. The process was discovered independently by two people in 1886: American chemist Charles Martin Hall (1863–1914) and Paul Héroult, who also discovered the electric arc furnace for melting iron and steel. In the Hall-Héroult process, molten aluminium forms at one electrode and sinks to the bottom of the vessel where it is collected. Oxygen is produced at the other electrode where it reacts with the carbon to form carbon monoxide and carbon dioxide gases. The molten cryolite remains unchanged and so more alumina can be added as required.

The ubiquitous tin

The metallic element tin was not just an essential component of some of the alloys needed to manufacture munitions for the First World War, it also had another ubiquitous use; in fact, so much so that the word became part of the everyday vocabulary of troops in the frontline. The troops, for instance, opened tins, or tin cans as they are also known, with a tin opener and cooked the contents in mess-tins.

During the war, the ubiquitous 'tin' was, as it still is today, most widely associated with food, as can be seen from the many references to tins and food in the literature of the First World War. For instance, German conscript Erich Maria Remarque (1898–1970), refers to a mess-tin on the very first page of his novel, *All Quiet on the Western Front*, based on his experiences in the frontline: 'We were even able to fill up a mess-tin for later, every one of us, and there are double rations of sausage and bread as well – that will keep us going'.[60]

The book *Christmas in the Trenches* provides a number of examples of how soldiers regarded tinned food as a godsend.[61] The book by Alan Wakefield, a curator at the Imperial War Museum, London, draws on several first-hand accounts by British troops of Christmas in the frontline. One soldier serving in Mesopotamia during Christmas 1917 describes doing 'pretty well for food issue rations'.[62] The rations included tinned kippers and a quarter of a tin of a type of meat and vegetable stew known as 'maconochie'. The soldier also purchased tins of salmon, sardines and cocoa from the company stores. Three years earlier, in the Christmas truce of 1914 between British and German troops on the Western

Front, some German soldiers gave British troops cigars in exchange for tins of bully beef.

The provision of adequate supplies of food was a critical issue during the First World War for the armies and civilian populations of the belligerent nations. Lack of food, starvation and malnutrition had a major impact on civilian morale and the outcome of the war, most notably in Germany and Russia. There are many accounts of civilians in these countries begging for food.

In one such account, Alfred Hall, a British territorial soldier captured by the Germans and put to work on a farm in Germany, comments in a diary entry for 3 March 1917 that the food question in Germany at that time was more important than the submarine question.[63] He describes a man and a woman wandering from farm to farm 'begging for bread'. In an entry for 12 September 1917, he writes: 'The starving appearance of the population of the small town of Siegen is very marked and one cannot help but have pity for the poor kids'.

In another account, Florence Farmborough (1887–1978), an English nurse on the Eastern Front, describes mobs in Russian cities shouting 'Peace and Bread!'[64] On 4 March 1917, she writes: 'Now that food has grown scarce in Petrograd and Moscow, disorder takes the shape of riots and insurrections ... They are aware that the war is at the root of their hardships'.

Inadequate supplies of often poor and monotonous food not only affected civilian populations during the war but also severely dented the morale of troops on the battlefield. The issue of food, for example, was one of several factors that triggered the French Army mutinies of 1917 on the Western Front in Northern France.[65] The importance of providing adequate supplies of food for the French Army was recognised over a century earlier by French general and Emperor of France Napoleon I (1769–1821). The saying 'an army marches on its stomach' is widely attributed to him.[66]

Napoleon, however, learnt the hard way when his army invaded Russia in June 1812. On 7 September, it defeated the Russian Army at Borodino, a village near Moscow, and a week later the French Army entered the city. However, the Russians had adopted a scorched-earth policy and other tactics to deprive the French troops of the food and forage that they had not already plundered en route to Moscow. The French Army consequently suffered from hunger during their occupation of Moscow and also fell foul of the Russian winter. The army, weakened by food shortages, was eventually forced to retreat and escape from Russia.

Napoleon's army might well have benefited from the new food preservation technologies, most notably canning, that were in their infancy at that time. Most raw foods are associated with microorganisms such as bacteria and fungi that thrive on them; some of these are harmless when consumed with the food, while others cause illness. Microorganisms can also result in food spoilage. Before the Industrial Revolution, food preservation traditionally relied on techniques such as salting, pickling, smoking over a fire, storing in concentrated sugar solutions, or drying in the sun. All these techniques involved the removal of water from the foods in order to reduce microbial activity.

In 1795, the French Directory, a group of executive directors of the French government, offered a 12,000 franc cash award to anyone who could develop a new and inexpensive technique for preserving the large quantities of food needed by the French armies. The prize was won in 1810, two years before the French invasion of Russia, by Parisian chef, distiller and confectioner Nicolas Appert (1749–1841).[67] Appert, who became known as the 'father of canning', demonstrated that the deterioration of foods could be slowed down by packing the foods into glass jars, sealing them with a cork and wax so they were airtight, and then immersing them in boiling water. The process, not understood at the time, sterilised the food by killing off the microorganisms. It was not until many years later that French chemist Louis Pasteur (1822–1895), the father of modern bacteriology, showed that microorganisms were responsible for food spoilage.

Appert was not the first to preserve food in this way, however. In 1808, Englishman Thomas Saddington was awarded a prize of five guineas by the London-based Royal Society of Arts for preserving food using a similar technique.[68] However, whereas Appert used his technique to preserve milk, meat, eggs, vegetables and cooked foods, Saddington only applied the technique to fruit.

In the same year that Appert won his prize, English merchant Peter Durand patented a method for preserving foods in containers made of glass, pottery, tin and other metals. Durand did not exploit his patent, but in 1812 sold it to the firm of Donkin, Hall and Gamble. The firm subsequently established a factory in Bermondsey, London, for preserving foods in sealed airtight cans made of tin-plated wrought iron. The products were mostly sold to the British Army and Royal Navy, and it was not until 1830 that tin cans containing tomatoes, peas, sardines and other foods first appeared on grocers' shelves. Over the following decades the sale of tinned foods rocketed.

With the development of techniques for the mass production of inexpensive steel in the latter half of the nineteenth century, steel began to replace wrought iron in the manufacture of tin cans. Tin, unlike iron and mild steel, resists corrosion and therefore acts as a protective coating. However, any damage to a tin can that exposes the steel to the atmosphere and moisture results in rapid corrosion.

The tin cans used in the First World War were fabricated from thin sheets of mild steel coated with a fine layer of tin. The tinning process involved dipping the steel sheets in molten tin, while, nowadays, electroplating is used to manufacture tin plate. One electrode is made of tin and the other is the steel sheet to be plated. Both electrodes are immersed in a bath containing an aqueous solution of the salt tin sulfate and connected to each other by an external electrical circuit. When the electricity is switched on, tin migrates from the tin electrode through the tin salt solution and deposits on the other electrode, that is the steel sheet.

Although electroplating was known in the nineteenth century, it was not until the 1930s that the process became well established.

Tin cans not only were used extensively in the First World War to provide food for troops, but they also had many other uses. Smoking, with the aid of the ubiquitous tin, was widely practiced during the war and provided much comfort, according to numerous accounts. Remarque, in his novel *All Quiet on the Western Front*, also focuses on the importance of tobacco, including cigars, cigarettes and chewing tobacco: 'Where would the soldier be without tobacco?'[69] Perhaps surprisingly, some soldiers not only survived the war but also survived a lifetime of smoking: 'I smoked from the time I was sixteen until I was a hundred and four', commented Fred Lloyd (1898–2005), a private in the British Army who served on the Western Front for two years. [70]

Tin containers were used for many other purposes as well. The German stick grenade of 1915, for instance, consisted of a tin cylinder attached to a wooden handle with a time fuse.[71] The cylinder contained the explosive and detonator. In the same year, British frontline troops made improvised bombs using empty jam tins filled with explosive – the so-called 'jam tin' bombs.

Fuses, detonators, tubes, primers, rockets, tracer shells and some types of cartridge were packed into tin cylinders with the lids soldered on to protect them from damp.[72] Tin containers were also used for storing and carrying oil and petrol, as well as for shipping a variety of chemicals. The shipping containers for the phosphorus used to make incendiary bombs, for example, typically consisted of sealed tin cans.[27] The phosphorus was submerged in water in the cans and the cans packed into wooden boxes. Sodium, a metal that reacts vigorously with water, was packed directly into hermetically sealed tin cans, or immersed in oil and stored in glass bottles, that were then placed in wooden boxes.

Millions upon millions of tins were manufactured, filled and distributed during the war. We should note, however, that not all First World War 'tin' items were made of the metal. For example, Brodie helmets, which, as we saw above were made of steel, were sometimes referred to as 'tin helmets' or 'tin hats'. Over the 1914 Christmas period, gift tins were distributed to over 2.6 million British soldiers and sailors serving at the time, and were known as Princess Mary gift tins. However, the 'tins' were actually made of brass. They contained tobacco and cigarettes, while non-smokers received 'tins' containing sweets

How the images of war depended on another precious metal

The first section of this chapter described how platinum, a precious metal, was essential for the production of the explosives used in the First World War. Another precious metal was also to play an important role in the war: silver, a metallic element that occurs naturally in its native form and in minerals such as argentite – a form of the inorganic compound silver sulfide. The rapid advances in both monochrome (black-and-white) and colour photography that occurred in the nineteenth century, and which were exploited so effectively in various ways during the First World War, were underpinned by the chemical properties of inorganic compounds containing this prized metallic element.

As early as 1727, German scientist Johann Schulze (1687–1744) had demonstrated that silver salts such as silver chloride and silver nitrate darken in the presence of light. Silver chloride along with silver bromide and silver iodide are known as silver halides because they all contain an element known as a halogen: chlorine, bromine and iodine respectively. The silver halides are white or yellowish compounds that are all insoluble in water. Most importantly they are photosensitive, so when they are exposed to light, a photochemical process occurs that breaks the silver halide up into dark metallic silver and the halogen.

Silver halides are prepared in their pure form by mixing an aqueous solution of silver nitrate, obtained by the reaction of silver with nitric acid, with an aqueous solution of a soluble halide salt such as sodium chloride or potassium bromide. On mixing, the insoluble silver halide salt precipitates out of the solution as a solid which can be filtered off, washed with water to remove impurities and dried.

The nineteenth century saw rapid advances in photography. For example, in 1829 French photographic pioneer Louis Daguerre (1789–1851) started working with French chemist Nicéphore Niépce (1765–1833) to develop an efficient method for producing permanent photographic images. Niépce had employed pewter plates coated with a solution of bitumen to capture the images. Pewter is an alloy consisting mainly of tin and small amounts of lead, copper or other metals.

Daguerre perfected Niépce's process four years after Niépce's death in what became known as the daguerreotype process. Made public in 1839, it employed a highly polished copper plate coated with silver iodide. The plate was placed in a camera and exposed to take the photographic image. After exposure the plate was developed with mercury vapour, which converted the coating on the portions of the plate exposed to light to a silver/mercury amalgam. The silver iodide remaining in the unexposed portions of the plate was then washed off. The process produced a high-quality positive image which, however, could not be reproduced.

The negative-positive photographic process was invented by Englishman William Henry Fox Talbot (1800–1877). In 1841, he patented a process that produced negative images on thin writing paper impregnated with silver iodide. The negatives, known as 'calotypes', were placed on top of paper photosensitised with silver chloride and then exposed to sunlight. The process created positive contact prints, known as 'salt prints'.

Daguerre and Fox Talbot are regarded as the founding fathers of photography, although others made significant contributions during the nineteenth century. As early as 1819, English scientist Sir John Herschel (1792–1871) had shown that the water-insoluble silver halides actually dissolve in aqueous solutions of sodium thiosulfate, an inorganic compound that he called 'hyposulphite of soda'. In 1839, he revealed in a lecture to the Royal Society in London that this 'hypo' solution could be used to fix the metallic silver image on photographic paper and at the same time wash away the unaffected silver halides which would otherwise cloud the image. Herschel is credited with introducing the word 'photography' in a letter to Talbot that year and coining the use of the photographic terms 'negative' and 'positive'.[73]

A major breakthrough occurred in 1851 when Englishman Frederick Scott Archer (1813–1857) introduced a photographic process for producing negatives that eventually made the daguerreotype and calotype obsolete. The wet collodion process, as it was called, employed a glass plate evenly coated with a mixture of potassium iodide and commercially available collodion dissolved in ether or ethanol.[74] Collodion is a form of nitrocellulose that was also used to make gelignite. The glass plate was then dipped into an aqueous solution of silver nitrate which converted the potassium iodide into photosensitive silver iodide. The sensitised wet plate was placed in the camera, and the photo was taken and developed to produce the negative before the plate was dry.

Twenty years later, another Englishman, Richard Leach Maddox (1816–1902), invented the gelatine dry plate, which consisted of a thin glass plate coated with a photosensitive emulsion of gelatine and a silver salt such as silver bromide. The plate could be used when dry and was therefore easier to use than the wet plate. Within ten years, factory-produced dry plates became commercially available. Maddox's introduction of the gelatine silver process effectively launched the photographic industry.

Over the following decades the dry plates were largely replaced by flexible, transparent celluloid roll film which had been introduced by American inventor and entrepreneur George Eastman (1854–1932) in the mid-1880s. The film consisted of an emulsion of a silver halide and gelatine on a transparent and flexible celluloid base. The celluloid, which was highly flammable, was prepared by plasticising nitrocellulose with the organic chemical camphor. In 1888, Eastman began to mass market the Kodak box camera that could take 'snapshots' on the roll film. Four years later, he founded the New York-based Eastman Kodak Company and in 1900 launched the Brownie roll film camera, various versions of which were produced until the 1960s.

Although the first colour photograph was produced by Scottish physicist James Clerk Maxwell (1831–1879) in 1861, it was not until 1907 that colour photography was established commercially. A glass plate was coated with a photosensitive silver halide emulsion and dyed grains of potato starch that acted as colour filters. The process, known as 'autochrome', was launched on the market by two French chemists who manufactured photographic materials – brothers Auguste (1862–1954) and Louis Lumière (1864–1948).

In the 1890s, the Lumière brothers developed a cine camera, the *cinématographe*, publicly screening their first motion picture films in Paris in 1895. From then on, the film industry developed rapidly. In the years leading up to the First World War, numerous silent motion pictures were made and shown in cinemas throughout Europe and the United States. The first dramatic feature-length film in colour, *The World, the Flesh and the Devil*, was shown in 1914. Now lost, the film employed the 'Kinemacolour' colour motion picture process invented by English film pioneer George Albert Smith (1864–1959). In this process, monochrome film was projected through rotating red and green filters.

Early motion pictures, whether monochrome or colour, relied on the same chemistry as that used for still photographs. A series of still shots, or frames, were photographed rapidly using roll film consisting of a celluloid base coated with a silver halide emulsion. The film was developed and the frames projected in rapid succession onto a screen to create the illusion of movement. By the start of the First World War, black-and-white photography and techniques for producing motion pictures had matured sufficiently to find widespread practical use among the warring nations. Colour photography, on the other hand, was still in its infancy and little used for war purposes.[75]

The history of war photography can be traced back to 1846 when Scottish photographers David Hill (1802–1870) and Robert Adamson (1821–1848) used the calotype process to take photographs of the Gordon Highlanders, an infantry regiment in the British Army, in Edinburgh Castle.[76] In 1855, English photographer Roger Fenton (1819–1869) sailed to the Crimea, now part of the Ukraine, and took more than 350 photographs of troops and campaign conditions in the Crimean War between Russia and Turkey (1853–1856). Britain and France had entered the war on the side of Turkey the year before. Fenton, the world's first accredited war photographer, was commissioned by Manchester publisher Thomas Agnew (1794–1871).

American Mathew Brady (1823–1896), sometimes regarded as the father of photojournalism, was one of the first to use a camera on the battlefield. He took photographs at the front early on in the American Civil War (1861–1865). Brady was soon joined by other famous American war photographers such as Alexander Gardner (1821–1882) and Timothy O'Sullivan (1840–1882). During the civil war, the Union Army of the Potomac alone issued passes to over 300 photographers eager to take dramatic shots of the war for newspapers and magazines.[77]

The following decades witnessed a rapid increase in war photography around the world. However, it was not until the end of the nineteenth century that motion picture films of war began to appear. The film and video archive at the Imperial War Museum in London, for example, holds moving-image material dating back to 1899, filmed during the Second Boer War (1899–1902) in South Africa.

In the First World War, newspapers and magazines illustrated their reports of the war with photographs submitted by war photographers and servicemen.

One such magazine was the weekly British magazine *Illustrated War News*, the first issue of which was published in August 1914. The use of cameras at the front, however, was restricted and photographs and films that might prove of value to the enemy were censored.

Photography and film were extensively used in the war for recruitment and training; for official records of military personnel, battles, and events such as victory parades; and for propaganda purposes to raise public feeling against the enemy and boost public support for the war effort. Some official documentaries of major battles, for example the Battle of the Somme in 1916, were not entirely accurate records of history as they contained some reconstructed scenes. Likewise, many propaganda photographs published in British war magazines such as the *War Illustrated* were clearly staged and of dubious authenticity. One such photo published in the magazine shows a German soldier robbing a fallen Russian soldier. It was almost certainly posed and intended as anti-German propaganda.[78]

The camera provided a record of military activities both abroad and also at home. In 1917, for instance, the surgeon-general at the Royal Victoria Hospital at Netley, near Southampton, arranged to have a film made about soldiers suffering from shell shock at the hospital who were under the care of two eminent Royal Army Medical Corps (RAMC) neurologists. The film aimed to demonstrate 'the efficacy of treatments at the hospital', explains Southampton-born writer Philip Hoare in a book about the hospital.[79] The twenty-seven-minute silent film was shot in black-and-white by British Pathé newsreel cameras.[80] It showed eighteen British soldiers, their symptoms and their treatment at the hospital in 1917, and their recovery in 1918 at the Seale Hayne Military Hospital, near Newton Abbot in Devon.

The hospital, built in 1856 to nurse wounded troops returning from the Crimean War, was used extensively throughout the First World War. For example, some of the 15,000 casualties of the 90,000-strong BEF that had sailed from Southampton and other ports in August 1914, the month that Britain declared war on Germany, were admitted to the hospital the following month. The hospital and an adjoining Red Cross hutted hospital at Netley treated some 50,000 casualties during the war.

The First World War triggered a number of advances in the use of photography, most notably in the application of aerial photography for mapping and reconnaissance of enemy positions and movements. The history of aerial photography dates back to 1858 when French balloonist and photographer Gaspar Felix Tournachon (1858–1910), known as 'Nadar', took photographs of houses from a tethered balloon. By the start of the war, cameras had been developed specifically for aerial reconnaissance and throughout the war aerial photography was widely used for military purposes by all sides. In the final year of the war, for example, French reconnaissance units were developing and printing up to 10,000 photographic plates a night during some battles.[81]

Soldiers were also fond of having their photographs taken in uniform, often with their fellow soldiers. In some cases the soldiers also used small pocket cameras to take unofficial snaps at the front. Their private official or unofficial photographs were often sent or taken home along with postcards showing local scenes near the front, the preparations for battle and the aftermath of battle. The captions on the postcards were sometimes censored so that the location of the scene could not be identified.

These personal photographs and postcards, together with the propaganda pictures and films, the aerial photographs, the documentary, training and recruitment films, the photographs published in newspapers and magazines, and the news footage shown at home, provide an enduring, vivid and sometimes horrifying pictorial record of war on a scale that had not been seen before. This extensive visual record of the First World War would not have been possible without the advances in photography that took place during the nineteenth and early twentieth centuries. Above all, they would not have been possible without a family of inorganic chemical compounds with unique photochemical properties containing a precious metal that is extracted from the earth, namely silver.

The new photography

In 1895, a scientific discovery that relied on the chemical and physical properties of a variety of metals and their compounds, including the silver halides, elicited intense excitement and interest among the scientific and medical professions and also much press coverage in newspapers.

The discovery was called 'a feat sensational enough and likely to stimulate even the uneducated imagination'. It was dubbed 'the new photography' by the *British Medical Journal* as it enabled 'hidden structures' to be photographed.[82] The new photography became known as radiography. 'The new photography would probably render great services in surgery', the journal predicted. And so it proved.

By the start of the First World War, radiography was being used extensively in the treatment of wounded soldiers. For example, one of the illustrations in Bland-Sutton's lecture on 'The value of radiography in the diagnosis of bullet wounds', published in December 1914, shows a radiograph, an X-ray, of 'a bullet lodged under the head of the fourth metatarsal bone, which it had broken'.[46] The only evidence of the injury to the naked eye was a red patch on the heel of the foot. Even earlier, on 16 October 1914, the Electro-Therapeutical Section of the London-based Royal Society of Medicine devoted a whole session to a 'Discussion on the localisation of foreign bodies (including bullets, &c.) by means of X-rays'.[83] The same section held a further 'Discussion on new methods for localisation of foreign bodies' on 15 January 1915.[84]

The military use of radiography dates back to March 1896 when the Anglo-Egyptian army, commanded by Horatio Herbert Kitchener (1850–1896), launched an expedition up the Nile which led to the conquest of Sudan two years later. The army equipment included two X-ray machines to help locate bullets and diagnose fractures. Remarkably, that expedition set out just five months after German physicist Wilhelm Konrad Röntgen had discovered X-rays. Röntgen named the radiation 'X-rays' as he wasn't sure what they were. Others called the radiation 'Röntgen rays'; the German word for X-ray was and still is *Röntgen*.

In November 1895, Röntgen had noticed that the discharge of electricity between two metal electrodes in a glass tube containing a gas at low pressure generated radiation that escaped out of the tube. It passed through a sheet of black cardboard and caused crystals of a chemical on a nearby table to glow in a darkened room. The chemical was barium platinocyanide, an inorganic compound containing the metallic elements barium and platinum, and the non-metallic elements carbon and nitrogen.

Röntgen subsequently demonstrated that the rays passed through the soft tissue of a human hand onto a screen coated with the compound and created an image of the hand. The physicist soon showed that photographic plates coated with a silver bromide gelatine emulsion were sensitive to X-rays as well as light and could therefore be used to record negative images of the hands. He then produced positive photographic prints from the X-ray plates.

The images in the plates and prints consisted of light areas and shadows, the darkness of which depended on the thickness and density of the soft tissue and bones. Soft tissue appeared as shadows and bones as light areas in the negative images, whereas they showed up as light and dark areas respectively in the positive prints. Metal objects such as wedding rings were distinctly visible in both the plates and prints.

Within months of Röntgen's discovery, scientists discovered that the radiation caused another inorganic chemical, calcium tungstate, to glow much more intensely than barium platinocyanide. Calcium tungstate contains oxygen and the metallic elements calcium and tungsten. By the end of 1896, some fifty or so books and pamphlets, and around 1,000 papers, had been published on X-rays and their applications. Over the next few years, the use of photographic glass plates to record X-ray images became standard and proved ever more popular.[73]

In 1901, Röntgen became the first ever winner of the Nobel Prize in Physics. He was awarded the prize 'in recognition of the extraordinary services he has rendered by the discovery of the remarkable rays subsequently named after him'.

Before the First World War, Belgium produced most of the glass needed to make X-ray plates. The war put a stop to the supply and celluloid photographic film was subsequently increasingly employed to record X-ray images.

One of the keenest advocates of the use of radiography in the war was another Nobel Prize winner, the Polish-born French physicist Marie Curie (1867–1934). In 1903, she was awarded the Nobel Prize in Physics jointly with two other

French physicists, her husband Pierre Curie (1859–1906) and Henri Becquerel (1852–1908) for their work on radioactivity.

In 1896, Becquerel discovered radioactivity while carrying out research on uranium salts derived from a uranium oxide-containing ore known as pitchblende or uraninite. Uranium is a radioactive silvery metallic element. Becquerel showed that the salts of the metal caused photographic plates to 'fog' even when covered by a layer of black paper. In 1898, the two Curies demonstrated that pitchblende and chalcolite, another uranium-rich mineral, contained two other radioactive metals which they named radium and poloniuim. They also coined the word 'radio-active' in the title of their scientific paper publishing their work on polonium.[85] Over the next four years, the Curies refined a vast amount of pitchblende and finally isolated a small quantity of a radioactive salt containing radium. The salt was pure radium chloride.

Four years later, in 1906, Pierre Curie was run over and killed by a two-horse cart in the *Rue Dauphine* in Paris.[86] Marie Curie continued her work on polonium and radium and in 1910 succeeded in extracting pure radium metal from radium chloride. She did so by the electrolysis of radium chloride in mercury and the removal of the mercury from the resulting amalgam of the two metals by distillation. The following year, Curie was awarded the Nobel Prize in Chemistry in recognition, according to the citation, 'of her services to the advancement of chemistry by the discovery of the elements radium and polonium, by the isolation of radium and the study of the nature and compounds of this remarkable element'. Curie was never able to isolate polonium as it is too unstable.

When war broke out in 1914, radiography was still a fledgling technique and there was little provision for it in military hospitals. Curie, assisted by her 17-year-old daughter Irène (1897–1956), began to organize X-ray services to help medical staff to locate shrapnel, bullets and broken bones in wounded soldiers.[87] She set up some twenty mobile vans, known as '*petite Curies*', that carried portable but primitive X-ray machines. The machines used radon, a radioactive gas produced by the radioactive decay of radium, as a source of radiation. Curie also arranged for X-ray units to be established in battlefield hospitals.

The year after Marie Curie's death in 1934, her daughter and son-in-law, Irène and Frédéric Joliot-Curie, jointly won the Nobel Prize in Chemistry in 1935 'in recognition of their synthesis of new radioactive elements'. They had carried out their research at the Radium Institute, now the Curie Institute, in Paris, which Marie Curie had established in 1914.

One of the pioneers of radiography during the First World War was British doctor Florence Stoney (1870–1932). When war broke out, she already had thirteen years' experience of X-ray treatment in English hospitals.[88] In September 1914, Stoney and her colleagues converted the abandoned Philharmonic Hall in

Antwerp, Belgium, into a hospital with X-ray facilities. Within days, the hospital's 135 beds were full of wounded troops. In October of that year, however, the city was bombarded by the Germans and the hospital was evacuated with the loss of its equipment.[89]

The following month, the team established a hospital in the sixteenth-century castle Château Tourlaville at Cherbourg, France. They managed to obtain funding and new equipment to run the hospital, although the facilities were basic: 'The sanitation was primitive, a pail system being necessary', remarked Stoney and fellow doctor Mabel Ramsay in a report on the hospital published in June 1915, adding: 'The greatest hardship has been the lack of water taps, except on the ground floor, and the lack of gas for heating'.[90] The electricity needed to run the X-ray machines was generated by water power in the grounds of the hospital. Between 8 November 1914, when the hospital received its first patients and the closure of the hospital on 24 March 1915, Stoney and her staff of six other female doctors and twelve nurses treated 120 critical wounded soldiers, of whom only ten died. In March 1915, Stoney was appointed as head of the X-ray department of Fulham Military Hospital, near London: 'More than 15,000 cases passed through her hands, many patients being sent from other hospitals for the localisation of bullets and pieces of shrapnel', her obituary in the *British Medical Journal* notes.[88]

X-ray technology advanced markedly during the war. In the early days of radiography, a radiograph required hours of exposure to the X-rays, so C. Thurstan Holland, a captain in the RAMC, explained in an address delivered to the Medical Students' Debating Society at the University of Liverpool in March 1917.[91] As apparatus improved, results became better and better and 'exposures were cut down from hours to minutes, minutes to seconds, and from seconds to fractions of a second', he said. 'Now it is possible to take radiographs so rapidly that a bullet in the heart – moving the whole time – can be taken so fast that its shadow is quite sharp; and it is even possible to take a series of stomach radiographs so quickly that one can cinematograph the stomach movements.'

6

Gas! GAS! Quick, boys!

Eyes like glowing coals

In January 1915, the German artillery fired thousands of tear gas shells at Russian troops on the Eastern Front. Although the attack employed a chemical warfare agent, it is not generally considered to mark the beginning of modern chemical warfare. That dubious distinction is normally accorded to the German chlorine attack at Ypres on 22 April 1915. The chlorine attack not only demonstrated that gas could be deployed in such a way as to kill and harm enemy troops on a mass scale, but also provided the first example in the history of warfare of the use of a weapon of mass destruction.

The attack and the subsequent gas attacks that occurred throughout the war caused immense suffering. The distress of troops caught in these attacks is dramatically portrayed in the art and literature of the time. The painting *Gassed* by American painter John Singer Sargent (1856–1925), for example, shows a line of British troops with bandages around their eyes heading towards a dressing station. Similarly, in her autobiography *Testament of Youth*, English writer Vera Brittain (1893–1970) vividly describes the early stages of mustard gas poisoning on troops in a ward where she was serving as a Voluntary Aid Detachment (VAD) nurse.[1] The 'poor things', poisoned during the Battle of Cambrai in November and December 1917, were 'burnt and blistered all over with great mustard-coloured suppurating blisters', she observes. Their blind eyes were temporarily or permanently sticky and stuck together, and they were 'always fighting for breath, with voices a mere whisper, saying that their throats are closing and they know they will choke'. Brittain refers to these toxic gases as 'inventions of the Devil'.

Numerous other reports document the horrendous effects of gas poisoning suffered by troops on all sides throughout the war. They include the following account by a German lance corporal who had enrolled as a volunteer in the 16th Bavarian Reserve Infantry Regiment in August 1914. In mid-October 1918, 'the British opened an attack with gas on the front south of Ypres … They used the yellow gas whose effect was unknown to us, at least from personal experience. I was destined to experience it that very night'.

The offensive described by the soldier, who served as a despatch runner in the regiment, took place on a hill south of Wervicq on the border between northern France and southern Belgium during the Battle of Courtrai:

> We were subjected for several hours to a heavy bombardment with gas bombs, which continued throughout the night with more or less intensity … About midnight a number of us were put out of action, some for ever. Towards morning I also began to feel pain. It increased with every quarter of an hour; and about seven o'clock my eyes were scorching as I staggered back and delivered the last despatch I was destined to carry in this war. A few hours later my eyes were like glowing coals and all was darkness around me.

The soldier, none other than the future German dictator Adolf Hitler (1889–1945), was sent to a hospital in the German town of Pasewalk for treatment. The description of the gas attack comes from James Murphy's 1939 English translation of Hitler's autobiography and political testament *Mein Kampf* (My Struggle) published in 1925.[2]

Hitler's hospital and military records do not specify the gas used in the attack, but merely indicate that he was lightly wounded and suffered from 'gas poisoning'. It has been widely assumed that because the Germans marked mustard gas shells with a yellow cross, the 'yellow gas' mentioned by Hitler was mustard gas. However, British psychologist David Lewis has obtained evidence that Hitler was possibly exposed to British 'White Star' gas, a mixture of chlorine, a yellowish-green gas, and phosgene during the attack, rather than mustard gas.[3] Furthermore, Lewis points out that Hitler was treated at a clinic specialising in nervous disorders where doctors 'concluded that his blindness was due to hysteria' rather than gas poisoning.

The Armistice was declared while Hitler was recuperating at the hospital in Pasewalk: 'For my part I then decided that I would take up political work', he notes in *Mein Kampf*. It is interesting to note that Hitler did not use chemical weapons during the Second World War.

Chemical warfare agents

Since the First World War, chemical warfare has been defined in various ways. British journalist Rob Evans, in his book *Gassed* (2000), defines it as 'the use of chemicals against the enemy to inflict toxic damage of varying degrees on man, animals or plants'.[4] It also embraces the development of defences against such chemicals.

Two years later, author Eric Croddy, writing about chemical and biological warfare, similarly defines chemical warfare weapons as 'poisons that kill, injure, or incapacitate'.[5] He specifically excludes incendiaries from the definition.

Augustin Prentiss (1915–2009), a lieutenant colonel in the Chemical Warfare Service of the United State Army, expresses a different view in *Chemicals in War – A Treatise on Chemical Warfare*.[6] He defines chemical agents as chemical substances used in combat and notes that they fall into three distinct groups: gases, smokes and incendiaries. Amos Fries, who was appointed Chief of the United States Army's Chemical Warfare Service in 1920, took a similar line: 'Chemical warfare includes, gas, smoke and incendiary materials and they can't well be subdivided'.[7]

Chemical warfare in its broadest sense has a long history. In 2009, British archaeologist Simon James produced evidence of what he believes was 'the first known use of chemical warfare.'[8] In AD 256, a Persian army besieged a Roman garrison in Dura-Europos, a city in Syria. The Romans intercepted a Persian siege mine with their own countermine. As they broke through, the Persians set fire to jars of bitumen and sulfur that not only caused the Roman tunnel to collapse but also produced clouds of choking gases that suffocated the Roman attackers.

With the growth of the chemical industry in the nineteenth century, there were several suggestions and proposals to develop and use chemical warfare agents. For example, Lyon Playfair (1818–1898), an eminent Scottish chemistry professor, advocated the deployment of cyanide- and phosphorus-filled shells against the Russians in the Crimean War between Russia and Turkey (1853–1856).[9] The British War Office and British Admiralty rejected the idea, although the British government did consult English chemist and physicist Michael Faraday (1791–1867) on the possibility of using poisonous chemicals against the Russians. Faraday, who is most famous for his pioneering experiments on electricity and magnetism, retorted that although it might be possible to employ them for this purpose, he would not participate in the venture as he thought it was barbaric.[10]

The possibility of using chemical agents as weapons of mass destruction caused growing alarm in the latter half of the nineteenth century. At an international peace conference held at The Hague in the Netherlands in 1899, signatories to an international treaty known as The Hague Convention of 1899 agreed 'to abstain from the use of all projectiles, the sole object of which is the diffusion of asphyxiating or deleterious gases'. The treaty effectively banned the use of shells filled with chemical agents even though at the time such shells had yet to be developed.

The declaration failed to stop Germany, France, Britain and other signatories to the convention, and the follow-up Hague Convention of 1907, from using chemical weapons in the First World War. The Germans carried out their first chlorine attack in April 1915 at the Second Battle of Ypres, with the British following up with their first chlorine attack at the Battle of Loos in September 1915. Chemical warfare using chlorine, phosgene, mustard gas and other chemical agents was waged not just on the Western Front, but on the Russian and Italian fronts during the war.

In his book on chemical warfare, Prentiss lists 'the principal chemical agents used in the war'.[11] They comprised forty-six gases, thirteen smoke agents and nine incendiary agents. In terms of the quantities produced and used, the most important of the gases were chlorine, phosgene, disphosgene, mustard gas, chloropicrin and the cyanides. Later in his book, Prentiss notes that Germany, France and England manufactured 68,100 tons, 36,955 tons, and 25,735 tons of 'battle gases' during the war respectively, whereas the United States, Austria, Italy, and Russia together manufactured 19,210 tons.[12] Overall, a total of 150,000 tons of these gases were produced, of which 25,000 tons were 'left on hand unused' at the end of the war. Another source puts the total amount of war gas produced for offensive use in the First World War at a staggering 176, 200 tons.[13]

By the start of the war, chemical warfare agents such as chlorine and phosgene had become widely used industrial chemicals with well-known toxicities. Some of these agents were released from cylinders in clouds that floated towards the enemy lines if the weather allowed, but most were delivered in around 65 million artillery chemical shells. That works out at less than 5 per cent of the estimated 1,400 million artillery shells fired throughout the war. More than 13 million chemical shells proved to be duds and another 13 million were not used and left behind at the end of the war.[14]

The chemical agents used in the war are often grouped into different classes according to their physiological effect on troops:

Lachrymators – often referred to as tear gases, irritate the eyes. The organic compound xylyl bromide is an example.

Sternutators – irritate the nasal passages causing sneezing, nausea, vomiting and headaches. Diphenylchlorarsine, an organic compound containing chlorine and arsenic, is an example.

Choking agents – also known as asphyxiators or lung irritants, irritate and injure the nose, throat, lungs and other parts of the respiratory tract. Chlorine and phosgene are well-known examples.

Vesicants cause blisters on exposed skin and also attack the eyes (resulting in temporary blindness), throats and lungs. They are sometimes called blister agents. Mustard gas is the most famous example.

Blood agents prevent blood from carrying oxygen. Cyanides are examples.

Chemical weapons were not just intended to kill, they also had other purposes, not least to harass and terrorise the enemy. The weapons inflicted misery on opposing troops and reduced their efficiency for combat. They confused and disoriented the opponents, inhibited their ability to communicate and slowed them down. Gas masks, when worn, were not only uncomfortable, but also impaired vision and caused extreme fatigue. And by poisoning enemy troops en masse, the weapons ensured that the enemy had to devote time and medical resources to the care of personnel suffering from the toxic effects of the gases.

Chemical warfare agents were therefore sometimes simply classified as either lethal gases or irritants: 'Gases employed in warfare may be divided into those actually lethal in their effects and those employed chiefly to clear the enemy from a position, or to inflict such temporary damage as will put the individual out of action for a period', remarked British industrial chemist Francis H. Carr in a lecture at the British Pharmaceutical Conference on 22 July 1919.[15] He cited chlorine, cyanides and phosgene as examples of lethal gases, classifying lachrymators, sternutators and vesicants as irritants.

The Germans colour-coded many of the chemical agents they used according to their physiological impact. Artillery shells marked with a white cross (in German: *Weisskreuz*) contained lachrymatory agents; a Blue Cross (*Blaukreuz*) shell was filled with an arsenic-containing sternutator, which affected the upper respiratory tract; a Green Cross (*Grünkreuz*) shell contained a choking agent such as phosgene or diphosgene; and a Yellow Cross (*Gelbkreuz*) shell was filled with a vesicant such as mustard gas.

Chemical warfare agents were dispersed not just as gases in the First World War, but also as solids and liquids. Mustard gas, for example, is an oily liquid that releases a toxic vapour. Diphenylchlorarsine is a white crystalline solid that melts at 45°C. When a shell containing the solid exploded, the agent was dispersed as a cloud of either solid particles or liquid droplets depending on the temperature of the cloud.[16]

The pros and cons of chemical warfare

Whereas the use of chemical weapons is now regarded with abhorrence, chemical warfare had many advocates during and following the First World War: 'Considering its power it has no equal', observed Amos Fries, '… the gas cloud is inescapable. It sweeps over and into everything in its path. No trench is too deep for it, no dugout, unless hermetically sealed, is safe from it. Night and darkness only heighten its effect. It is the only weapon that is as effective in a fog or the inky blackness of a moonless night as in the most brilliant sunshine.'[7]

Some even considered that the use of chemicals weapons was preferable to the use of high-explosive shells. Prentiss, for example, noted that 'in general, gas causes less suffering than wounds from other weapons.'[17] Others suggested that the paralysis and demoralisation of the enemy troops were as effective as and more humane than slaying them.[18] Yet not everyone at the time was happy with the use of gas. In his lecture, Carr commented that British investigators undertook work on these gases 'with sickened feelings of revolt against such debasement of science'. [15]

Following the German chlorine attack at Ypres in April 1915, the Allies responded by rapidly developing and using their own chemical weapons. In May 1915, for example, the British established four Special Companies of the Royal Engineers, later to become known as the Special Brigade, to take responsibility for waging chemical warfare. The companies, under the command of Lieutenant Colonel Charles Howard Foulkes (1875–1969), included science graduates and industrial chemists. Within five months, they mounted their first British gas attack delivering chlorine from cylinders at the Battle of Loos. In order to keep the use of gas secret, chlorine was referred to as 'the accessory'.[19]

According to military historian Gordon Corrigan, the British soon became the leaders in gas warfare and 'the Germans never regained their early advantage'.[20] By the end of the war, Allied production of toxic agents and chemical munitions was matching that of the Germans. For example, in November 1918, the month of the Armistice, a United States Army site in Maryland, known as Edgewood Arsenal, had the capacity to fill over 2.5 million gas shells each month.[21]

'Chemical warfare developed from a novelty in 1915 to an approved component of the British method of waging war by 1916', observes historian Albert Palazzo in his book on the subject published in 2000.[22] Furthermore, 'it made a significant contribution to the eventual British victory in 1918', gas becoming 'the most effective method' of neutralising the enemy's guns and achieving fire supremacy. It also lowered the morale of the enemy and reduced their ability to resist.

'Gas, in general, and phosgene in particular, was much more effective as a psychological weapon than as a tactical weapon', according to four British experts on phosgene and its use in war.[23] However, unlike Palazzo, they conclude that chemical warfare did not appear to have a decisive impact upon the outcome of the war or indeed on many battles. The use of gas was an effective way of removing men from the battlefield, but 'did not prove to be an effective way of killing the enemy'.

The last point is borne out by the casualty and fatality statistics of the war even allowing for the fact that such statistics are approximate and vary widely depending on their source. Poison gas resulted in a total of 1.5 million recorded casualties among all the belligerent armies, but caused less than 1 per cent of all fatal battle casualties, reports historian Roy MacLeod in a study published in 1998.[24]

In his book on the gas attacks at Ypres in 1915, published in 2009, John Lee, an expert on the history of the First World War, puts a lower but quite specific figure on the number gas casualties: 'There were a total of 1,286,853 gas casualties in all the armies of the First World War – about 75 per cent caused by gas shell and the rest by gas cloud … Of this total, 93 per cent survived the experience, although 12 per cent would have some sort of permanent disability'.[25] Thus the remaining

7 per cent or some 90,000 of the men affected by gas died. In 1937 Prentiss produced similar statistics, stating that the total number of gas casualties in the war was 1,296,853, exactly 10,000 more than the number quoted by Lee.[26]

MacLeod, on the other hand, notes that there were 'up to' 90,000 deaths in all. He observes that the number almost certainly underestimates the large number of gas fatalities in the Russian armies, in the Balkans and Mesopotamia, and those in the West 'who died subsequently from respiratory illnesses and other complications experienced after war records were compiled'.

MacLeod adds that 'best estimates' suggest that of the 2.9 million British casualties in the war, artillery shelling accounted for 58 per cent and machine-gun and rifle fire for 39 per cent. Although British gas casualties rose from 7.2 per cent of all wounded in 1917 to 15 per cent in 1918, the percentage of gas fatalities dropped from 3.4 per cent to 2.9 per cent of all battlefield deaths over the same period. In comparison, the French suffered some 130,000 gas casualties, of which 17.5 per cent were fatal, and the Germans 107,000, of which 5 per cent were fatal. According to Prentiss, of the 72,807 American gas casualties in the war, 71,345 received non-fatal injuries and 1,462, around 2 per cent, died.[26] Another American source reports that 72,000 or about 25 per cent of a total of 272,000 US casualties in the war were gas casualties.[27] Approximately 1,200 of the gas casualties proved fatal, dying either in hospital or in action due to gas exposure.

All these statistics show that poison gas accounted for a very low percentage of the total battlefield casualties and fatalities on all sides during the First World War. When it came to maiming and killing, bullets, shrapnel shells and high-explosive shells were far more effective in numerical terms.

Tear gases

A dozen or more organic chemicals were employed as tear gases, also known as lachrymators, by the combatant armies in the First World War. As these chemicals were used as non-lethal irritants rather than lethal agents, none of the belligerent nations considered them to contravene the Hague Convention of 1899.

The concept of using tear gases against enemy troops dates back to the nineteenth century when Scottish chemist William Ramsay (1852–1916) carried out experiments on acrolein, a flammable liquid with a pungent smell. The organic chemical, also known as 2-propenal or acrylic aldehyde, was produced by the decomposition of the glycerine formed when tallow, a form of animal fat or grease, was burnt.

Ramsay is more famous for the discovery, along with other scientists, of the chemical elements argon in 1894, helium in 1895, and krypton, neon and xenon in 1898. These elements, which are renowned for their lack of chemical reactivity, are known as inert or noble gases. Ramsay is the only person to have discovered an entire group of elements in the Periodic Table of Chemical Elements. In 1904, he was awarded the Nobel Prize in Chemistry 'in recognition of his services in the discovery of the inert gaseous elements in air, and his determination of their place in the periodic system'.

Following his experiments on acrolein, Ramsay suggested to the British War Office that the compound could be employed as a tear gas against an enemy, although they decided not to use it.[28] The French, on the other hand, introduced the chemical in January 1916 as a filling for gas grenades and artillery gas shells.[29] It was not a great success, however, as the material is chemically unstable in air. Acrolein is unique among the lachrymators used in the First World War in that it does not contain chlorine, bromine or iodine. These three chemical elements, known as halogens, form part of the group of elements next to the group of inert gases in the periodic table.

The first tear gas, and indeed the first chemical warfare agent, to be used in the war was an organic chemical containing bromine. The chemical, a colourless liquid with a pungent odour, was ethyl bromoacetate. Two British organic chemists, William Henry Perkin (1838–1907) and Baldwin Francis Duppa (1828–1873), reported the preparation of the compound in a paper published in a German chemistry journal in 1858.[30] They prepared the compound by treating acetic acid with bromine and then combining the product, bromoacetic acid, with ethanol.

According to some reports, the French police used ethyl bromoacetate as tear gas as early as 1912 to help them capture criminals.[31] Ludwig Haber (1921–2004), the son of Fritz Haber who developed an efficient way of synthesising ammonia, disputes such reports in his book on chemical warfare in the First World War.[32] He notes that two chemists at the municipal laboratory in Paris had recommended the use of the chemical agent to the police for riot control, but the police did not use it. Two years later, however, the French did use ethyl bromoacetate as tear gas. In August 1914, the month Germany declared war on France, the French Army began to employ hand and rifle grenades filled with the liquid. By November, however, supplies of the bromine needed to make the compound had begun to dry up. The French therefore replaced the tear gas with chloroacetone, another lachrymatory halogen-containing organic compound.

In June the following year, the Germans introduced the lachrymatory chlorine-containing chemical agent, chloromethyl chloroformate – otherwise known as 'K-Stoff' or 'C-Stoff'. In low concentrations the compound acted as a tear gas, but exposure at higher concentrations was lethal. A month later, the Germans started to use bromoacetone, also known as 'B-Stoff', as a tear gas.[33] The compound has a similar chemical structure to chloroacetone, the only difference being that the chlorine is replaced by bromine. The Germans also employed three other bromine-containing organic compounds as tear gases: xylyl bromide and benzyl bromide, both known as 'T-Stoff', and bromomethylethyl ketone or 'Bn-Stoff'. They were introduced in January, March and July 1915 respectively.

Over the same period, the French turned their attention to iodine-containing organic chemicals as lachrymatory agents. They developed iodoacetone which, like bromoacetone, is a chemical relation of chloroacetone. The army first fired artillery shells containing iodoacetone in August 1915. By November, a growing shortage of the acetone needed to make the compound had forced the French to switch to benzyl iodide as a tear gas. Just two months later, they replaced it with acrolein.

Meanwhile, in Britain, Ramsay was working with the War Office to develop chemical agents for shells and bombs.[34] He suggested that aerial bombs might be filled with the blood agent hydrogen cyanide, more commonly known at the time as prussic acid. The office rejected the proposal as it contravened the 1899 Hague Convention. Ramsay then consulted with two English chemists, Herbert Brereton Baker (1862–1935) and Jocelyn Field Thorpe (1872–1940), who tested some fifty compounds as potential tear gases in a trench at Imperial College, London.

They settled on ethyl iodoacetate, a colourless oily liquid, to fill grenades, shells and bombs. The compound was not new, however, as British chemists Perkin and Duppa had described its preparation over fifty years earlier in 1859.[35] Unlike some of the other compounds tested, the iodoacetate did not corrode cast iron and steel, so munitions and canisters containing the liquid could therefore be transported and stored without fear of leaks. Furthermore, the iodine needed to manufacture the chemical was readily available in Britain, whereas the bromine needed to make ethyl bromoacetate was in short supply.

Ethyl iodoacetate was called 'SK' after South Kensington, the location of Imperial College, and became the standard British lachrymatory agent. The army used cast iron shells filled with the chemical at the Battle of Loos in September 1915, and it was subsequently used to fill gas grenades and Stokes mortar bombs.

Towards the end of the war, the Americans developed a chlorine-containing organic compound, chloroacetophenone, code-named 'CN', as a lachrymatory agent. The compound was discovered by German chemist Carl Gräbe (1841–1927) in 1869 but not used in the war. In 1965, the chemical was introduced as a riot control agent in a product known as 'Chemical Mace'. As the compound is a solid, it was dissolved in a volatile organic solvent such as butanol and dispersed as an aerosol of liquid droplets from a spray can.

Sternutators

According to Prentiss, a total of 7,315 tons of sternutators, otherwise known as sneeze gases, were manufactured during the war. Germany produced 7,200 tons of this total, whereas England and France produced relatively small amounts: 100 tons and 15 tons respectively.[12]

The Germans fired their Blue Cross shells containing these respiratory irritants simultaneously with the Green Cross shells filled with lethal choking agents. The sternutators were intended to penetrate the gas masks that protected enemy troops from the lethal gases and induce sneezing, nausea and vomiting. The troops would consequently be forced to remove their masks and therefore expose themselves to the choking gases.

That was the theory, but in practice the dispersion of these chemicals in a suitable form proved difficult. Diphenylchlorarsine, first synthesized in 1881, introduced by the Germans in July 1917 and used in the Spring Offensive of 1918 against the Allies on the Western Front, was a typical example. The chemical, like most of the other sternutators used by the Germans, was an arsenic-containing organic compound known as an arsenical.

Prentiss reports that the Germans filled some 14 million Blue Cross artillery shells with diphenylchlorarsine and its close chemical cousin diphenylcyanarsine.[36] The latter was introduced in May 1918. Both arsenicals are colourless crystalline solids and powerful irritants to the upper respiratory tract. When inhaled, they cause not only sneezing but also intense pain in the chest, shortness of breath, retching and vomiting.

The Germans intended these solid chemicals to generate a dust that would not be trapped by the absorbent chemicals in the Allies' gas masks. However, the particles produced when the shells burst were far too big to penetrate the masks. Furthermore, for maximum impact, the shells should have exploded just above the ground where the Allied troops were standing, but they were not fused to do so. The failure to find a way of dispersing the sternutators as a mask-penetrating dust 'stands out as one of the few technical mistakes in chemical warfare that the Germans made during the World War'.[16]

Green and lethal

The title of this book, 'Gas! GAS! Quick, boys!' comes from 'Dulce et Decorum Est', one of the finest and most famous poems by British war poet Wilfred Owen (1893–1918). Owen wrote the poem during the last year of his life – he was killed in action on 4 November 1918, a week before the end of war. The final two lines of the poem are:

> The old Lie: *Dulce et decorum est*
> *Pro patria mori.*

The italicised Latin, meaning it is sweet and fitting to die for one's country, is taken from a poem by Roman poet and satirist Horace (65–68 BC). In his poem, Owen describes the vision of a soldier dying from gas poisoning. Through a 'thick green light' cast by a cloud of chlorine gas, Owen sees him yelling, stumbling, floundering and drowning 'as under a green sea'.

Chlorine, a member of the halogen group of elements, is one of just a few coloured gases. It is a dense, sharp-smelling gas variously described as greenish-yellow or yellow-green in colour. Its name derives from the Greek word *chloros* which means pale green. The element dissolves in water to form chlorine water, a slightly acidic solution that not only contains dissolved chlorine but also low concentrations of hypochlorous and hydrochloric acids, and has antiseptic and bleaching properties. Solutions that contain low concentrations of chlorine-releasing chemicals, such as sodium hypochlorite, are widely used to sterilise drinking water and purify swimming pools.

When chlorine gas is inhaled in high concentrations it can prove fatal. In lower concentrations it attacks the respiratory tract, causing coughing, nausea, vomiting, dizziness and headaches, and can result in pulmonary oedema – the accumulation of fluids in the lungs.

Most importantly for chemical warfare, the gas is over twice as heavy as air. When it was released and carried by the wind, it clung to the ground and sank into trenches. On the other hand, it was also a non-persistent gas in air: after reaching its target area and gassing the opposing troops, the gas rapidly dispersed to safe levels. Attacking troops could then approach and enter the area without fear of being gassed themselves.

The green gaseous element was first prepared and studied by German-born Swedish chemist Carl Wilhelm Scheele (1742–1786) in 1774. He made it by adding hydrochloric acid, then known as muriatic acid, to manganese dioxide, an inorganic compound that occurs naturally as the mineral pyrolusite. The compound contains two atoms of oxygen for every atom of manganese. Scheele and other chemists at the time mistakenly thought the green gas was a compound that combined the acid with the oxygen extracted from the manganese dioxide. They therefore called it oxymuriatic acid. It was not until 1810 that English chemist Humphry Davy (1778–1829) showed that the gas did not contain oxygen and that it was an element, naming it chlorine after its colour.

Chlorine was originally produced by the reaction of hydrochloric acid and pyrolusite. However, by the end of the nineteenth century the element was principally manufactured by the electrolysis of a solution of sodium chloride in water, also known as brine. When electricity is passed through the solution, chlorine is generated at one of the electrodes and hydrogen at the other. The process leaves a solution of sodium hydroxide, a caustic soda solution, in the electrolytic cell.

In the years running up to the First World War, Germany produced large quantities of chlorine for the manufacture of synthetic dyes, pharmaceuticals and other chlorine-containing organic chemical products. The country was therefore well equipped not only to produce chlorine as a war gas, but also to manufacture organic chemical warfare agents that contained chlorine.

By the end of 1914, shell shortages and the stalemate on the Western Front had induced both the Germans and the Allies to consider filling shells and bombs with highly toxic chemical agents that could kill large numbers of opposing troops or at least drive them out of their trenches. The Germans had already attempted to boost the effectiveness of shrapnel shells by embedding the shrapnel balls in dianisidine chlorosulfonate, a sternutatory chlorine-containing organic chemical. However, the shells proved ineffective when used against the French at Neuve Chapelle in October 1914. In January 1915, the Germans fired shells filled with the tear gas xylyl bromide, the so-called T-Stoff shells, on the Eastern Front, but these were also ineffective.

A month earlier, Fritz Haber, the German chemist more famous for devising the synthesis of ammonia from atmospheric nitrogen, had come up with the idea of using chlorine as a weapon. He suggested to the German High Command that clouds of chlorine gas could be discharged from industrial canisters against soldiers in enemy trenches. The High Command liked his proposal and asked him to test the use of the gas as a weapon for trench warfare. It subsequently gave the go-ahead for a chlorine gas assault along a sector of the Western Front near Ypres as 'an experiment'. Haber recruited a team of some 500 gas pioneers, including a number of scientists. The team assembled at a site in Cologne where they practised discharging steel cylinders filled with chlorine under high pressure.[37] The scientist-soldier gas pioneers included four future Nobel Laureates: Haber himself, who was to win the Nobel Prize in Chemistry in 1918; the German chemist Otto Hahn (1879–1968), who won the 1944 Nobel Prize in Chemistry for his research on nuclear fission; and two physicists – the German-born American James Franck (1882–1964) and the German Gustav Hertz (1887–1975), who shared the 1925 Nobel Prize in Physics for their research on energy transfer between molecules. Haber's team soon expanded into a 1,600-strong gas Pioneer Regiment.

On 22 April 1915 the 'father of modern chemical warfare', as Haber became known, directed the German gas troops' infamous chlorine cloud assault against the Allies at Langemarck, near Ypres. The British first used chlorine, discharged from cylinders as clouds of gas, at the Battle of Loos on the Western Front in Northern France on 25 September 1915. Chlorine became the principal gas used by all sides for cloud gas operations on the Eastern, Italian and Western Fronts throughout the war.[38]

From 1916 onwards, the British, French, Russians and Germans employed chlorine mixed with other chemical agents in their cloud gas attacks. These agents were mainly phosgene, diphosgene or chloropicrin, although hydrogen sulfide, sulfur dichloride, stannic chloride (also known as tin tetrachloride) and arsenic trichloride were also used. The British primarily used chlorine-phosgene (codenamed White Star), chlorine-chloropicrin (Yellow Star), chlorine-sulfur dichloride (Blue Star) and chlorine-hydrogen sulfide (Green Star) mixtures. Pure chlorine was coded Red Star.

Phosgene

Humphry Davy, the English chemist who identified chlorine as an element, had a younger brother, John Davy (1790–1868), also a chemist. John Davy rose to fame for his discovery of one of the most dangerous gases to be used in the chemical warfare of the First World War. That gas was phosgene, also known as carbonyl chloride. From a chemical point of view, the compound is relatively simple. Each molecule of the gas contains just one atom each of carbon and oxygen, and two atoms of chlorine.

John Davy discovered the gas in 1811 by mixing a volume of the green chlorine gas with an equal volume of carbon monoxide, a colourless odourless gas. He observed that the colour of the mixture disappeared when it was exposed to sunlight. The resulting gas, when released into the atmosphere from its glass container, had an odour quite different to that of chlorine and was 'more intolerable and suffocating than chlorine itself'. Furthermore, the gas affected 'the eyes in a peculiar manner, producing a rapid flow of tears' and occasional painful sensations.

After carrying out numerous chemical tests on the gas, he concluded that it was a compound formed by the combination of carbon monoxide, or carbonic oxide as he called it, and chlorine. He named the compound 'phosgene' after the Greek for 'formed by light'. John Davy described his discovery in a paper entitled 'On a gaseous compound of carbonic oxide and chlorine', presented to the Royal Society of London in February 1812.[39] From the 1880s onwards, phosgene began to be used as an industrial chemical for the manufacture of dyes and other synthetic organic chemical products. By 1914 Germany was producing the material in 'moderately large quantities for various peaceful purposes'.[40] The country was therefore in a good position to exploit the gas as a chemical warfare agent.

The colourless gas is far more toxic and also more dense than chlorine.[41] It drifted down into enemy trenches and shell craters and had a not unpleasant smell of new-mown hay. Victims were therefore not always aware that they had inhaled the gas. Exposure to a single low dose could prove harmless apart from initial symptoms such as eye-watering, coughing and dizziness.[42] Moderate doses, however, usually had a delayed effect on the heart, which frequently proved fatal some twelve to twenty-four hours after the early symptoms had disappeared. A few breaths of the gas in high concentrations could result in death from heart failure within a few minutes.

The poison works by attacking proteins in the respiratory parts of the lungs, causing the transfer of oxygen to blood to be disrupted and fluid to accumulate in the cavities of the lungs. As a result, victims drowned in their own fluids. According to Prentiss, phosgene caused 80 per cent of all the gas fatalities in the First World War.[43]

The Germans first used the gas, which they called D-Stoff, in a mixture with chlorine in a cloud gas operation against the British on the Western

Front in Flanders on 19 December 1915. The less dense chlorine helped the phosgene to spread. The German gas unit discharged some 177 tons of the gaseous mixture from 9,300 cylinders, resulting in 1,069 British gas casualties, of whom 120 died.[44]

In France, soon after the German chlorine attack in April 1915, a group led by Victor Grignard (1871–1935), professor of organic chemistry at the University of Nancy, began to investigate the synthesis of phosgene. The chemists carried out their experiments in a chemistry laboratory at the Sorbonne in Paris.

Grignard is famous in the world of chemistry for discovering a class of organic compounds containing both magnesium and a halogen. These so-called Grignard reagents are employed in Grignard reactions, a type of reaction that is widely used by chemists to synthesise certain types of organic compound. Grignard jointly won the Nobel Prize in Chemistry in 1912 with another Frenchman, Paul Sabatier (1854–1941), who was a chemistry professor at the University of Toulouse. It was the first time that the Nobel Prize in Chemistry had been shared. Grignard was awarded the prize for his discovery of the reagents named after him, while Sabatier won the prize for his research on the use of finely-divided metals to speed up organic chemical reactions.

Grignard's group at the Sorbonne included several wounded chemists who had been sent home from the Western Front.[45] The group carried out research on the preparation of phosgene by the reaction of carbon tetrachloride and a solution of sulfur trioxide in sulfuric acid, commonly known as oleum or fuming sulfuric acid. The procedure was used to 'a limited extent' for the industrial manufacture of phosgene in France during the war.[46]

The French employed phosgene as an artillery-shell filler at Verdun in February 1916, and it was first used by the British at the Battle of the Somme in June 1916. Phosgene became, as Prentiss notes, 'the principal offensive battle gas of the Allies, being used in enormous quantities in cylinders, artillery shell, trench mortars, bombs, and projector drums'.[43]

Diphosgene

Diphosgene, like phosgene, is a choking agent and the Germans first used the chemical in May 1916 against the French at the Battle of Verdun. In one attack alone, the Germans fired around 100,000 diphosgene shells.[47] Both phosgene and diphosgene became the principal chemical agents throughout 1917 due to the superior effect on enemy troops.[48] Diphosgene, also known as trichloromethyl chloroformate, is a close relation to phosgene, although its chemical structure is more complicated. Each diphosgene molecule contains four atoms of the halogen chlorine, and two atoms each of carbon and oxygen.

Although the two chemicals have similar levels of toxicity as lethal lung irritants, they have important differences as chemical warfare agents. As we saw

Boys in Khaki. The author's uncle George Curtis (centre) served in the First World War.

German chemist Fritz Haber, known as the 'father of modern chemical warfare'. (Courtesy: Archiv der Max-Planck-Gesellschaft, Berlin-Dahlem)

Left: Ammonia synthesis. Haber in a laboratory at the University of Karlsruhe, Germany. His process was crucial for the German war effort. (Courtesy: Archiv der Max-Planck-Gesellschaft, Berlin-Dahlem)

Below: Carl Bosch, chemist and engineer who worked with Haber to develop an industrial process for making ammonia. (Courtesy: Archiv der Max-Planck-Gesellschaft, Berlin-Dahlem)

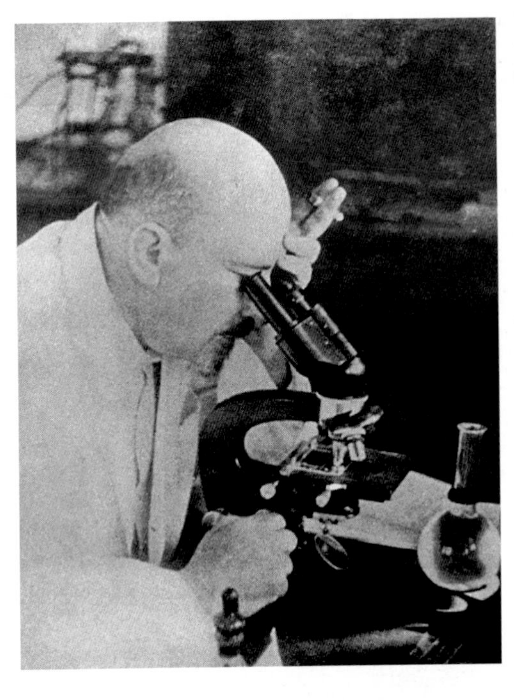

Chaim Weizmann. His industrial fermentation process played a key role in cordite production in Britain during the war. He later became the first President of the State of Israel. (Courtesy: Weizmann Institute of Science)

Paul Ehrlich is regarded as the founder of chemotherapy. He also synthesised a drug for treating syphilis that was widely used in the war. (Courtesy: Paul Ehrlich-Institute, Langen, Germany)

Martha Whiteley, a chemist at Imperial College London, assembled a team of female chemists to produce anaesthetics for the British war effort. (Courtesy: Archives Imperial College London)

Above: Opium poppies. Morphine, codeine and opium extracted from poppy pods were widely used as a painkillers in the war.

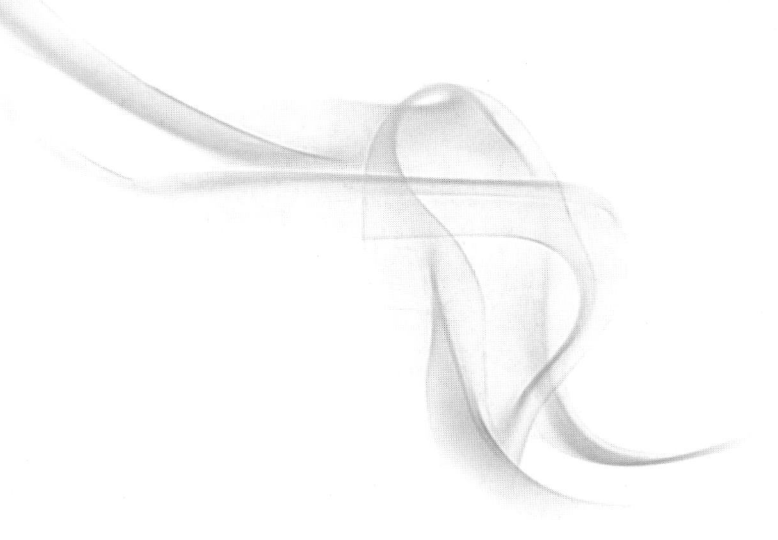

Opposite top: Lochnager crater, 90ft deep and 300ft in diameter, near Albert, France. It was created when a company of British Army pioneers exploded mines containing 24 tons of ammonal on the first day of the Somme Offensive, 1 July 1916.

Opposite bottom: Trenches at Sanctuary Wood, near Ypres.

St Mary's Hospital. The inoculation department of the London hospital produced ten million doses of typhoid vaccine during the war.

Royal Victoria Hospital. The chapel is all that remains of the hospital at Netley, near Southampton, which treated some 50,000 sick and wounded First World War servicemen.

Taken from a German newspaper, this image shows a German soldier undergoing oxygen treatment after exposure to poisonous gases. (*The Illustrated War News,* 14 July 1915)

'Germany's pride in her poisoning tactics'. Taken from an illustrated German newspaper, this drawing shows Canadian troops being forced to retreat at St Julien, north-east of Ypres, in the face of a poisonous gas cloud. (*The Illustrated War News,* 21 July 1915)

Above: British soldiers filling jam tin bombs with Turkish shell and barbed-wire scraps on the Gallipoli peninsular. (*The Illustrated War News,* 28 July 1915)

Opposite top: High-explosive shells being manufactured at the 'Krupp's of France', the Creusot munitions works. (*The Illustrated War News,* 28 July 1915)

Opposite below: High-explosive shells ready for the final process. The Creusot munitions works, France. (*The Illustrated War News,* 28 July 1915)

Early version of British respirator, 1915. (*The Illustrated War News,* 4 August 1915)

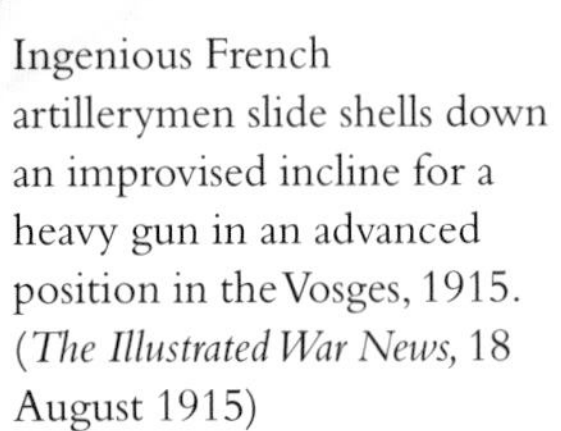

Ingenious French artillerymen slide shells down an improvised incline for a heavy gun in an advanced position in the Vosges, 1915. (*The Illustrated War News,* 18 August 1915)

A French Rimailho gun (short, 155mm, rapid-fire) engaged in range-regulating fire while subject to counter-battery fire from German artillery. (*The Illustrated War News,* 18 August 1915)

A French infantryman throwing a grenade. (*The Illustrated War News,* 25 August 1915)

'Every day demonstrates that the Great War is essentially one of science and munitions'. Bomb-making at a training school. (*The Illustrated War News,* 8 September 1915)

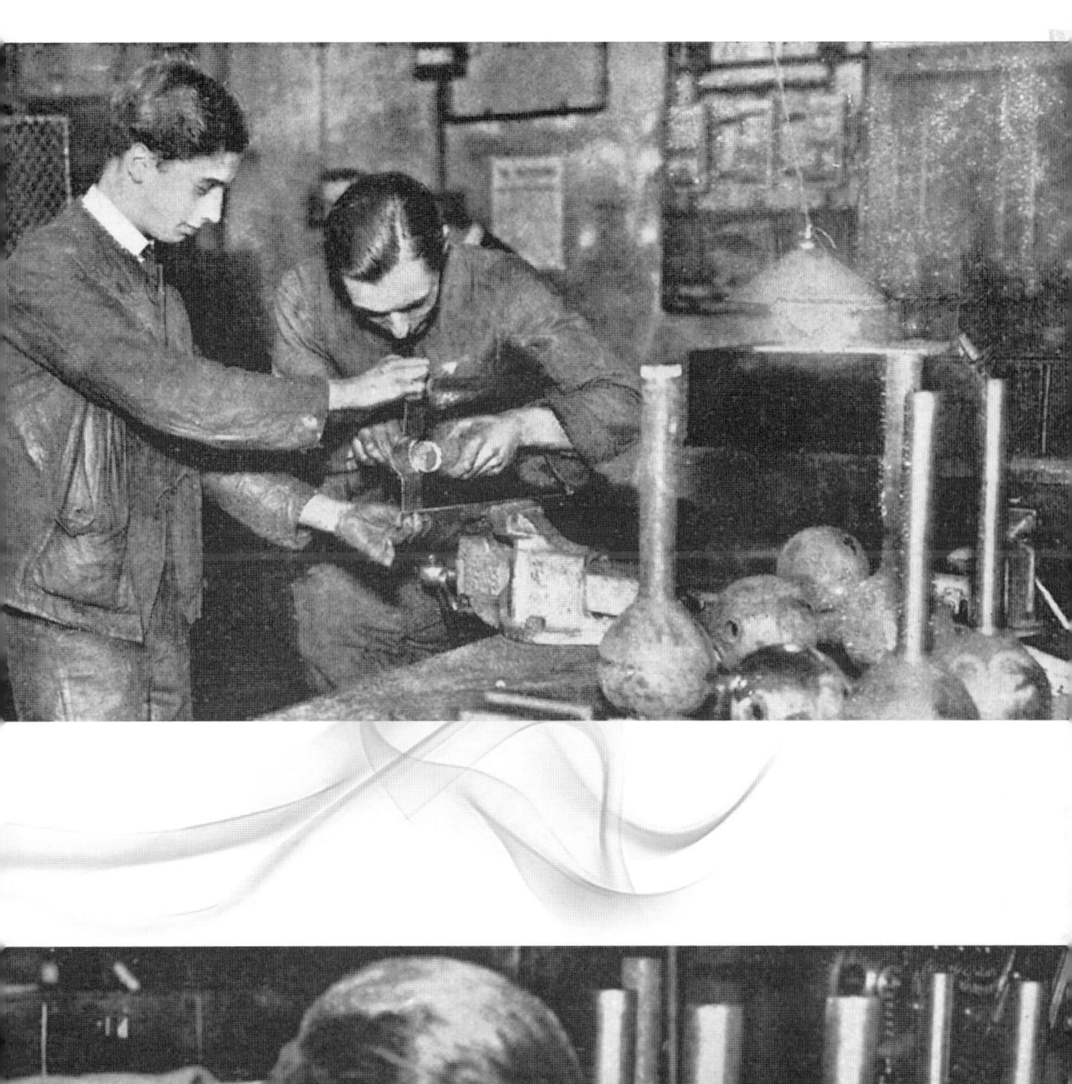

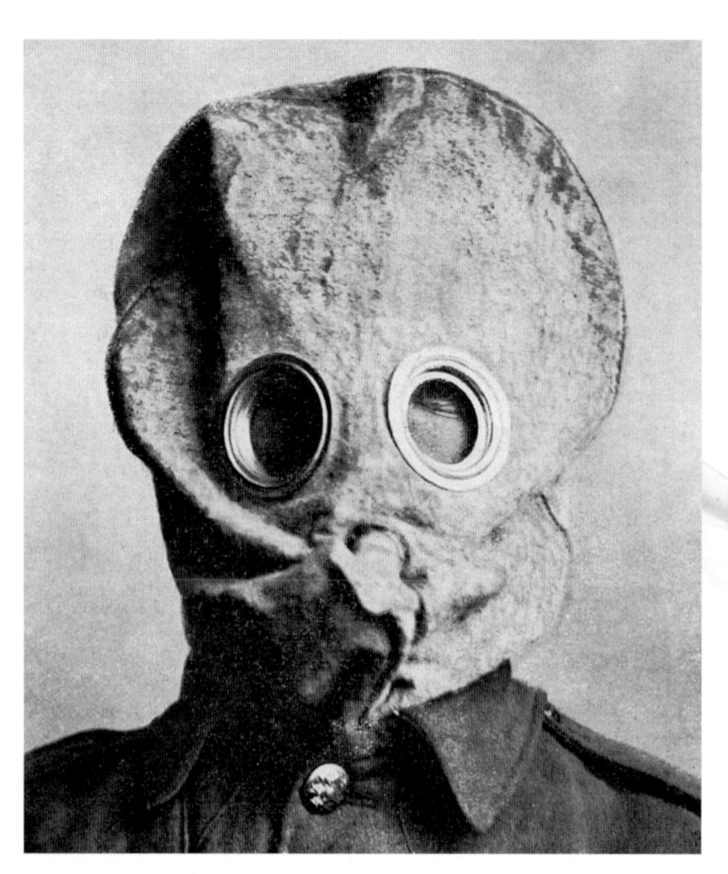

'The Modern Warrior: a British soldier in the new anti-gas helmet'. (*The Illustrated War News,* 29 September 1915)

French sappers in the crater of a mine explosion which has destroyed German trenches. (*The Illustrated War News,* 3 March 1915)

British heavy artillery. (*The Illustrated War News,* 24 March 1915)

'Hardly less fatal in effect than the flying fragments from bursting shells charged with high explosives are the fumes that are given off at the moment of explosion.' A British officer wearing a respirator to protect against 'shell gases'. The special respirator features a small valve which opens and shuts as the wearer breathes. (*The Illustrated War News,* 24 March 1915)

Steel-rolling mill, Krupp's gun factory, Germany. (*The Illustrated War News,* 7 April 1915)

Unexploded German artillery shell, weighing nearly a ton and measuring 5ft long, seen next to a French 75mm shell and a German 77mm shell. (*The Illustrated War News,* 14 April 1915)

Aerial photograph of French gas and flamethrower attack on German trenches in Flanders, 1 January 1917. (NARA)

above, a variety of delivery systems were used in the First World War to discharge phosgene. The chemical was, for example, mixed with chlorine and discharged from cylinders for cloud gas attacks. However, as its boiling point is just 8°C, it is a gas at room temperature under normal conditions. Therefore, in order to pour liquid phosgene into a cylinder, bomb, drum, or shell, it was necessary to keep the material refrigerated at less than 8°C.

Diphosgene, on the other hand, boils at 128°C and so the oily liquid was comparatively easy to pour into shells at room temperature. When the shells burst, they released a heavy vapour that clung to the ground for about thirty minutes. It was therefore more persistent than lethal gases such as phosgene or chlorine and was unsuitable for cloud gas operations. For these reasons, diphosgene became 'the principal *killing* gas used *in shells* during the war', Prentiss notes.[49]

The Germans manufactured diphosgene in the dye-manufacturing plants of two chemical companies: Bayer, located at Leverkusen, a city on the Rhine, and Höchst, based near Frankfurt.[50] The two companies used different chemical routes to synthesise diphosgene, but both routes involved two chemical steps. Bayer first produced methyl formate, a type of organic compound known as an ester, by the reaction of formic acid and methanol. Formic acid, also known as methanoic acid, occurs naturally in ants and stinging nettles, and is manufactured industrially from carbon monoxide. Likewise, methanol also occurs naturally and is manufactured from carbon monoxide. In the second step, Bayer bubbled chlorine through the methyl formate, a liquid, to chlorinate both the methyl and formate components of the compound and produce the trichloromethyl chloroformate (diphosgene). Höchst, on the other hand, first prepared methyl chloroformate, on oily liquid, by treating methanol with phosgene. The company then prepared the diphosgene by chlorinating the methyl group in methyl chloroformate.

From June 1915 until the end of the war, Bayer produced an average of 300 tons of the chemical per month, while Höchst averaged 139 tons per month starting September 1916.

Chloropicrin

> When an aqueous solution of nitropicric acid is poured into a retort containing a great excess of hypochlorite of lime, the mixture heats spontaneously, and an aromatic pungent vapour, which affects the eyes very powerfully, is immediately evolved. If heat is applied to the retort so soon as the mixture begins to boil, a very large quantity of a colourless heavy oil comes over along with the vapours of water, and condenses in the receiver.

So wrote Scottish organic chemist John Stenhouse (1809–1880) in 1848 following his discovery of a compound which he named 'chloropicrine'.[51] The compound is now generally known as 'chloropicrin' without the 'e' or even more simply as 'chlorpicrin'. Chloropicrin was one of many organic compounds that Stenhouse, one of the co-founders of the Chemical Society of London in 1841, discovered during his lifetime.[52]

The chemical structure of chloropicrin is closely related to that of chloroform, which contains one atom each of carbon and hydrogen, and three atoms of chlorine. Chloropicrin has a nitro group instead of the hydrogen atom. The compound is sometimes known as trichloronitromethane or more simply as nitrochloroform.

The 'nitropicric acid' Stenhouse refers to in his paper on chloropicrin was trinitrophenol, the yellow solid widely known as picric acid or lyddite that was used as a high-explosive bursting charge in shells during the First World War. The other starting material, 'hypochlorite of lime', is now known as calcium hypochlorite and is one of the components of bleaching powder.

Like picric acid, phosgene, and so many other chemicals discovered in the eighteenth and nineteenth centuries, chloropicrin was used extensively in the First World War. It is more toxic than chlorine, but less so than phosgene or diphosgene. The compound is a liquid at room temperature that releases a powerful lachrymatory vapour. The British called the chemical 'vomiting gas' as it injures the stomach and intestines causing nausea and vomiting. It also acted as a lethal asphyxiant.

The Germans, British, French, and Russians used chloropicrin in artillery shells and trench mortar bombs. Sometimes they mixed it with chlorine so that it could be delivered from cylinders for cloud gas operations. It had the particular advantage of being able to penetrate gas masks that protected troops from chlorine and phosgene.

Chloropicrin is a highly persistent gas that clings to the ground after it is discharged, with combatants finding that it could take several hours to disperse to safe levels depending on the weather conditions and the state of the ground. The chemical therefore had limited use for tactical offensive operations as it could also gas the attacking troops.

The Russians were the first to use the gas on the battlefield in August 1916. The gas was produced under the direction of the Commission for the Production of Asphyxiating Gases (CPAG) which had been set up by Russia's War Ministry.[53] The British had started conducting trials with the gas as early as February 1916, but they did not employ it for military operations until 1917.[54]

Blood agents

From 1916 onwards, Austria and the Allies, notably France, began to introduce highly volatile blood agents as a means of chemical warfare. When breathed in, these chemicals rapidly pass from the lungs into the blood stream and prevent the body's cells from using oxygen. Death from heart failure can occur within minutes, even when only traces of an agent are inhaled.

The most important blood agents used in the war were three chemically related compounds: hydrogen cyanide, cyanogen bromide and cyanogen chloride. A hydrogen cyanide molecule consists of one atom each of hydrogen, carbon and nitrogen, while molecules of the other two compounds contain a bromine and chlorine atom respectively in place of the hydrogen atom.

Hydrogen cyanide, the deadliest and most infamous of these agents, was first prepared in 1782 by Scheele, the same Swedish chemist who discovered chlorine. He prepared the cyanide, which he called prussic acid, by heating a synthetic blue pigment, a chemical combination of iron atoms and cyanide groups and known as 'Prussian Blue', with sulfuric acid. Hydrogen cyanide is a volatile, colourless liquid with a boiling point of just 26°C and smells faintly of bitter almonds. It was the active component of 'Zyklon B', a pesticide product that was used by Nazi Germany to kill prisoners in the death camps during the Second World War.

In 1916 the CPAG in Russia began to produce both hydrogen cyanide and cyanogen chloride as chemical warfare agents, according to Nathan Brooks of New Mexico State University, United States, who is an expert on the history of chemistry in Russia and the Soviet Union.[53]

The French used shells filled with hydrogen cyanide mixed with stannic chloride and chloroform in the Battle of the Somme in July 1916. Stannic chloride was also one of the chemicals the French mixed with chlorine for their cloud gas attacks. The chemical contains tin and chlorine in the ratio one to four. It is a lachrymatory colourless liquid at room temperature that reacts with the moisture in air to produce fumes, and was therefore sometimes used as a smoke agent.

The French also tried mixing hydrogen cyanide with chloropicrin or the inorganic compound arsenic trichloride. The latter, sometimes known as 'butter of arsenic', is an intensely poisonous, oily liquid that fumes in air and, like stannic chloride, was also mixed with chlorine for cloud gas operations.

The French called their hydrogen cyanide-containing mixtures 'Vincennite'. Both French and American troops fired Vincennite shells during the war. However, they proved ineffective as the chemical contents of the shells evaporated and dissipated far too rapidly: 'Vincennite is practically worthless and was not in the least feared by the enemy', remarked a chief gas officer in the American Expeditionary Force (AEF) in a report on gas attacks published just after the war.[55]

The Americans only used three types of gas shell in any number during the war: the Vincennite shell; a shell containing a mixture of phosgene and stannic chloride; and one containing mustard gas. They were all supplied by the French. 'Of the two effective types of filling, the phosgene shell is non-persistent and lethal, mustard gas persistent and vesicant', commented the AEF chief gas officer.

Mustard gas

In March 1917, British and French experts in chemical warfare attended the first Anglo-French Congress for Chemical Warfare, held in Paris. The main topics of discussion at the meeting concerned the problems of defence against chemical weapons such as chlorine, phosgene, hydrogen cyanide and the lachrymatory gases.

'The delegates had not been back in their laboratories many weeks before the Germans suddenly launched on the British forces the first attack with mustard gas', reports J. Enrique Zanetti, who represented the AEF at a follow-up meeting: the first Inter-Allied Congress for Chemical Warfare that also took place in Paris over three days in September 1917. Zanetti was to become a captain in the European Division of the Chemical Warfare Service the following year, his report on the congress appearing in 1919.[56] Mustard gas was such a formidable weapon, Zanetti remarks, that if the Germans had been able to produce it on the same large scale that the Allies had been manufacturing it at the time of the Armistice, 'there is little doubt that the Allied lines would have melted, to use the poet's words:

> Like snow on the sands
> Touched by the April wind.

In other words, according to Zanetti, mustard gas could have won the war for Germany, and 'No attempt was made to belittle the importance of the new arrival in chemical warfare'.

Mustard gas was the last of the lethal chemical warfare agents to be used on a large scale in the First World War. Chlorine was first used in April 1915, phosgene in December 1915, diphosgene in May 1916, and chloropicrin in August 1916. The Germans introduced mustard gas on the night of 12/13 July 1917 when they bombarded the British frontlines east of Ypres with some 50,000 mustard gas shells.[57] Over the next ten days, they fired over a million shells containing a total of some 2,500 tons of the gas.[58] Thereafter until the end of the war, the Germans used the chemical as the primary filling for their artillery gas shells.

The Allies soon identified the new agent and put their chemists to work to develop processes for its production. Within a year, the French had begun to manufacture mustard gas shells and first fired them on the Western Front on 16 June 1918. They also supplied the Americans with the shells, but as Brook observes, 'It appears that Russia never managed to produce mustard gas'.[53]

Following consultations with the French on the best process to use to manufacture the gas, the British eventually produced stocks of the chemical at a plant in Avonmouth, near Bristol, by August 1918.[59] The British first fired shells containing the agent on 29 September – the opening day of the Battle of

St Quentin Canal. The offensive, which also involved American and Australian troops, punched through the heavily fortified, in-depth German defences known as the Hindenburg Line that ran for some 75 miles along the central and northern sectors of the Western Front in northern France.

Mustard gas was perhaps the most dreadful of all the chemical warfare agents used in the First World War. The chemical, however, is not actually a gas, but an oily liquid that has to be heated to over 217°C to make it boil. At ambient temperatures, the liquid releases a 'weak, sweet, agreeable odor', states the *Merck Index*, an authoritative reference work covering thousands of synthetic and naturally-occurring inorganic and organic chemicals and chemical products.[60]

The thick viscous liquid was typically mixed with an organic solvent such as carbon tetrachloride before loading into shells.[61] The British used a ratio of four parts mustard gas to one part solvent. When a mustard gas shell exploded, it dispersed the chemical as an aerosol – a suspension of liquid droplets in air. Many soldiers were unable to detect the odour on the battlefield before the droplets penetrated their army uniforms, including leather and rubber boots, and attacked the skin. Furthermore, the liquid also released a vapour that could be inhaled. Full protection against mustard gas thus required not only gas masks but also head-to-toe protective suits. Such clothing was not widely available to frontline troops in 1918.

Later in the war, Grignard, the French Nobel Laureate who had worked on the synthesis of phosgene, developed a precise test for mustard gas.[62] One of the chemical names of the agent is dichloroethyl sulfide. Each molecule of the compound consists of a sulfur atom bonded to two chloroethyl groups, each of which contains a chlorine atom, two carbon atoms and four hydrogen atoms. Grignard's test employed sodium iodide, a white crystalline salt that is soluble in water. When the mustard gas came into contact with an aqueous solution of the salt, the iodine in the salt replaced the chlorine in the dichloroethyl group. The reaction precipitated diiodoethyl sulfide, an easily-detectable bright yellow crystalline solid that is insoluble in water. The test was able to detect one-hundredth of a gram of mustard gas in one cubic metre of air.

Mustard gas was one of the least volatile and therefore most persistent of the chemical agents used in the war, sometimes lying around on the ground for days on end. The persistency of chemical warfare agents such as mustard gas depended on a number of factors. Water, for example, destroyed the gas and other agents like phosgene by a process known as hydrolysis.

'The time which must elapse before unprotected troops can safely occupy ground which has been bombarded with chemical shell varies with different shells, weather conditions, and nature of the ground', explains the AEF chief gas officer: 'Sunshine, rain, and wind diminish the persistence while low temperature and absence of wind increase it. In dugouts or cellars, gas may persist for days if the ventilation is poor and, likewise, slightly volatile chemicals may remain undecomposed below the surface of the ground for a week or more unless the ground is disturbed by digging.'[55]

Taking a typical example of a gassed area swept by a 3mph wind, the officer recommended that troops should not enter such areas before the following approximate times had elapsed:

	Open ground	Woods
Vincennite mixture	8 minutes	30 minutes
Phosgene	20 minutes	3 hours
Chloropicrin	1 hour	20 hours
Ethyliodoacetate (tear gas)	6 hours	36 hours
Mustard gas	24 hours	7–10 days

During winter, mustard gas could persist for even longer periods, sometimes several weeks, both in open ground and in woods. The persistency limited the use of the gas as a tactical offensive weapon, but was useful as a tactical defensive weapon as it could be used to deter enemy troops from occupying an area. According to Hartcup, the Germans found its persistency particularly useful when they were forced to retreat from an area when an attack failed.[63]

A slow and painful death

The *Merck Index* aptly describes mustard gas as a 'deadly vesicant'. Exposure to the chemical results in the formation of very small fluid-containing blisters in the skin, known as vesicles, which eventually grow into much larger blisters. The chemical is also lethal, with a toxicity level similar to that of phosgene.

In the First World War, mustard gas temporarily blinded troops, irritated and burnt their skin, produced intense hoarseness and caused inflammation of the lungs. Victims also suffered from both internal and external bleeding. Most frightening of all, however, was that it was an insidious poison: troops exposed to the chemical did not experience any symptoms for several hours.

'Probably 90 per cent of the mustard gas casualties were due to skin burns; burns on the feet and legs when standing in or walking through areas shelled with the liquid, on the hands from picking up guns or other equipment, or from sitting or lying down on the ground', observes George Burrell, who headed the Research Division of the Chemical Warfare Service in the United States.[64]

Death could then be slow and painful for victims exposed to high levels of the liquid: 'Soldiers often took four to five weeks to die, putting a further load on the enemy's medical services and the pain was so bad that most soldiers had to be strapped to their beds', notes science historian Patrick Coffey.[65]

Mustard gas was not new to the First World War. It was synthesised in various ways from chlorine- and sulfur-containing compounds by French, British and

German chemists during the latter half of the nineteenth century and early twentieth century.

The chemical is also known as sulfur mustard to distinguish it from its cousin nitrogen mustard, which has a similar chemical structure to that of sulfur mustard but contains nitrogen rather than sulfur. Nitrogen mustard is also a vesicant but is more toxic than its sulfur counterpart. The nitrogen compound, a liquid with a faint fishy smell, was first produced as a potential chemical weapon in the 1920s and '30s.[66]

The Germans called sulfur mustard 'Yellow Cross' after the markings on the gas shell. They also used the codename 'Lost' for the chemical, which was derived from the first two letters of the surnames of the two chemists who developed a method for mass producing the agent for use in the war. They were Wilhelm Lommel, who was employed at the Bayer chemical works in Leverkusen, and Wilhelm Steinkopf (1879–1949), who headed the chemical weapons research team at the Kaiser Wilhelm Institute for Physical Chemistry and Electrochemistry in Berlin,

Fritz Haber, director of the institute, had examined the gas in 1916 and rejected its use as a chemical weapon because he considered its toxicity to be too low. Its vesicant properties, although known at the time, were not thought to be significant. The Germans only realised its power as a chemical weapon after they became aware of the heavy toll of casualties following its use at Ypres in July 1917.[57]

From mid-1917 to the end of the war, Bayer produced a total of 4,800 tons of mustard gas, averaging some 300 tons per month.[50] The company employed a fairly complicated procedure involving five steps to manufacture the chemical using ethanol, 'chloride of lime' which is chemically similar to calcium hypochlorite, carbon dioxide, sodium sulfide and hydrochloric acid.

The French and Russians called the product 'Yperite' as it was first employed at Ypres. The British codenamed the new German agent 'HS', which was short for 'Hun Stuff', and, along with the Americans, also called the chemical 'mustard gas' because its odour resembled that of mustard.

The king of the battle gases

In one sense, mustard gas was the worst of the chemical warfare agents used in the First World War. In terms of the pain and suffering it caused, it was the most horrific of the war gases. Yet in another sense it was the best or at least the most effective of the chemical weapons used in the war. For example, between July 1917 and the end of the war, German mustard gas accounted for around 125,000 British casualties, that is some 70 per cent of all British gas casualties in the war.[67] However only 1,859 or 1.5 per cent of the mustard gas casualties proved fatal.

Fritz Haber, after his initial reservations, considered mustard gas to be 'a fabulous success', and soon after it was first used, it became known as the 'king of the battle gases'.[68] Many years later, Prentiss came to the same conclusion: 'Of all the casualty gases used in the war, mustard gas was by far the most effective'.[11] A total of 12,000 tons of mustard gas were used in the First World War,[69] and there are also accounts of German aircraft dropping mustard gas in small balloons in gas attacks against AEF troops in March 1918.[70] However, the attacks did not result in any reported casualties and in all probability used relatively little of the agent. On the other hand, some 9 million artillery shells containing the vesicant were fired by Germany, Britain and France during the war, resulting in a total of around 400,000 casualties. According to Prentiss' calculations, that equated to one casualty for every 60lb of mustard gas used, or one for every twenty-two-and-a-half mustard gas shells fired. That compared with 500lb of high explosive for each of the 10 million high explosive casualties in the war, and with 5,000 rounds of rifle or machine-gun ammunition for each of the 10 million bullet-wound casualties.

Whereas sulfur mustard was an effective chemical weapon, its cousin nitrogen mustard proved to be effective in a completely different and beneficial way. Sulfur and nitrogen mustards are cytotoxic chemicals, that is they can damage or kill cells in the body. In the 1940s, researchers began to investigate the potential therapeutic value of these compounds and in particular their ability to destroy cancerous cells in lymph nodes and bone marrow. Nitrogen mustard subsequently became the first clinically useful anti-cancer cytotoxic drug and has since been used in various forms of chemotherapy to treat Hodgkin's disease, some types of leukaemia and other malignant diseases. The chemical works by interfering with the chemical mechanism in cancer cells that causes them to divide and proliferate uncontrollably. In the world of chemistry and medicine, it is known not only as nitrogen mustard but also by a variety of other names, including mustine, chlormethine and mechlorethamine.

Dew of death

Chlorvinyldichlorarsine is America's principal contribution to the *materia chemica* of the First World War, states Prentiss in his book on chemicals in war.[71] The arsenic- and chlorine-containing organic compound is an oily, colourless liquid when pure, with a faint smell of geraniums, and is a powerful lung irritant and vesicant. It is both a local and general poison of deadly strength, and, Prentiss believed, was 'undoubtedly superior' to any of the other First World War gases.[72] Many years later, the agent became known as 'America's World War I weapon of mass destruction'.[73]

The compound, an arsenical, was first synthesised by PhD student Julius Arthur Nieuwland (1878–1936) at the Catholic University of America in Washington DC. Nieuwland was carrying out research on the chemistry of acetylene, an organic compound which contains two atoms each of carbon and hydrogen in each molecule. In his doctoral thesis 'Some Reactions of Acetylene', published in 1904, he describes the reactions of acetylene with seventy-five other compounds.[74] One of these was arsenic trichloride, and in order to speed up the reaction he used aluminium chloride as a catalyst. The product was chlorvinyldichlorarsine. In his thesis, Nieuwland describes the material as a tarry substance that had a nauseating smell and was 'extremely poisonous'.

Nieuwland was born in Belgium to Flemish parents and while at the Catholic University studied botany and chemistry. He also completed his training for the priesthood, before being ordained in 1903. Following his PhD, he took up a post as professor of botany at the University of Notre Dame in Indiana and then, in 1918, became professor of organic chemistry. His research on the chemistry of acetylene eventually led to the synthesis of neoprene, also known as polychloroprene, the first commercial synthetic rubber.

After completing his PhD, Nieuwland did not continue his research on the highly toxic arsenical. However, in 1918, his thesis was drawn to the attention of American chemist Winford Lee Lewis (1878–1943), who was working on the development of arsenicals as chemical warfare agents at the Catholic University. Lewis saw that Nieuwland had studied the reaction of acetylene and arsenic trichloride and decided to carry out further investigations. He devised a procedure for preparing and purifying the arsenical and also determined its chemical structure. Lewis showed that each molecule of the compound contained two atoms each of carbon and hydrogen, as well as one arsenic atom and three chlorine atoms. The compound was named 'Lewisite' after him.

The Americans code-named the chemical 'G-34' and 'Methyl'. They manufactured it in secret using Lewis's procedure at a factory near Cleveland, Ohio. The AEF Gas Service Section then began to formulate plans to use aircraft to spray the chemical over German cities. In anticipation of this deadly precipitation from the skies, Amos Fries, who headed the Gas Service and subsequently the United States Army's Chemical Warfare Service, dubbed Lewisite 'the dew of death'. In November 1918, the United States shipped a batch of about 150 tons of the deathly dew to Europe. Before it arrived, however, the Armistice was signed and the batch was 'unceremoniously' dumped into the sea.[66]

Gas defence

Chemicals not only harm and kill, they also protect and prevent harm. This paradox is no more apparent than in the chemical warfare of the First World War. On the one hand, chemists on all sides were beavering away to generate more and more potent chemical warfare agents, yet on the other hand and at the same time, chemists were furiously attempting to find chemicals to counteract these agents.

The primary means of gas defence employed by the belligerent nations in the war were gas masks. In the context of chemical warfare, the terms 'gas mask' or simply 'mask' are often used synonymously with the term 'respirator'; however, some of the more sophisticated gas masks developed later in the war were actually helmets. Other gas masks comprised not only a mask that covered the face, but also a separate gas filtration canister containing protective chemicals that absorbed or chemically destroyed the toxic gases. The masks and helmets were typically made of fabrics impregnated with chemicals that protected the troops against the toxic gases. The canisters were either screwed into the masks, the so-called 'snout canister masks', or connected to the masks by tubes.

Although gas masks and respirators developed rapidly during the war, they were not new. In the 1850s and '60s for example, Stenhouse, the discoverer of chloropicrin, designed and patented wood charcoal air filters and respirators that employed wood charcoal as an absorbent material for removing noxious gases.[52]

Protection initially proved ineffective

In the early days of gas warfare, when chlorine first became a weapon of mass destruction, protection was crude and largely ineffective. There are accounts of troops covering their faces with wet handkerchiefs, socks, towels, wads of cotton, or bits of flannel, dampened with water or urine, to protect themselves from the gas.[75]

In theory, both these liquids should provide protection against chlorine gas. As we saw above, the gas dissolves in water to form a weakly acidic solution known as chlorine water, which is far less dangerous than chlorine gas. Similarly, the major component of urine, apart from water, is urea, a nitrogen-containing organic compound. An aqueous solution of urea reacts with chlorine and neutralises the acids in chlorine water in a complicated sequence of chemical steps, yielding a solution containing a variety of relatively safe chemical products such as nitrates and carbon dioxide.[76] When urine is left to stand, the dissolved urea reacts with water and breaks down into ammonia and carbon dioxide in a process known as hydrolysis. The process is catalysed by an enzyme called urease. Ammonia, an alkaline compound, neutralises chlorine to form the water-soluble compound ammonium chloride.

Within a couple of weeks of the first German chlorine gas attack on 22 April 1915, British troops were issued with fairly primitive respirators consisting of pads of cotton wool wrapped in muslin envelopes that covered their noses and mouths. They had been made by women in England in response to an appeal in the *Daily Mail* on 28 April 1915. 'We learned that handkerchiefs filled with earth and kept moist would keep some of the gas out, and by the time the first novelty had worn off we were receiving private respirators from England', British chemist Samuel J.M. Auld wrote in a book on chemical warfare published in 1918.[77] Auld, a former professor of agricultural chemistry at Reading University, England, was a major in the Royal Berkshire Regiment and a member of the British Military Mission to the United States.

On 17 January 1918, Auld delivered a lecture on gas warfare to the Washington Academy of Sciences in which he noted that troops soaked their respirator pads in aqueous solutions of sodium carbonate and sodium thiosulfate, although sometimes only water was available.[78] Other troops dipped their pads into sodium bicarbonate solutions during the chlorine gas attacks.

Sodium carbonate and sodium bicarbonate are commonly encountered as washing soda and baking soda respectively. Sodium thiosulfate solution, often referred to as 'hypo', was widely used as a fixing agent in the photographic process at the time of the war. All three sodium compounds dissolve in water to form alkaline solutions and so, by definition, neutralise acids. In respirator pads, the alkaline solutions were thus able to neutralise the hypochlorous and hydrochloric acids formed when chlorine gas dissolved in the solutions.

Pad respirators evolved rapidly following the first German gas attack: 'A new type appeared almost every week', observed Auld in his lecture in Washington DC. One such respirator was designed by Leslie F. Barley (1890–1979), a chemical advisor to the British Army. The pad was placed over the mouth and nose and fastened by a strap behind the head, and was known as the 'Barley mask'.[79]

Pad respirators such as the Barley mask required frequent soaking in the solutions that were available to the troops at the front. The pads proved to be almost totally useless, however. When dry, they allowed chlorine to pass through them, and when wet they tended to form an airtight seal over the nose and mouth, making it difficult or impossible to breath.

These improvised masks were soon abandoned in favour of what became known as 'Black Veil Respirators', which were similar to those used by the Germans. The respirator employed a length of black cotton netting, like that used to make mourning veils, folded to form a pocket. A mouthpad of cotton waste soaked in an aqueous solution of sodium thiosulfate, sodium carbonate, and glycerine, which prolonged the lifetime of the solution, was inserted into the pocket. The respirator was tied around the head with the pad over the mouth and nose, and air was able to pass through the cotton waste while the solution neutralised the chlorine. An eye flap attached to the veil could be pulled over the eyes for protection against tear gas.

These masks had their limitations, however. They only afforded protection against chlorine for a few minutes and they were not gastight. The 'Hypo helmet', designed by Canadian medical officer Cluny MacPherson (1879–1966) and developed by Barley and other British chemists in uniform, offered some improvement. The helmet, officially known as the 'British Smoke Hood', consisted of a flannel bag soaked in the thiosulfate-carbonate-glycerine solution used for Black Veil Respirators. A rectangular eyepiece made of mica, celluloid or cellulose acetate allowed the soldier to see through the hood. The soldier put the hood on by pulling it over his head and tucking the open end under the collar of his uniform.

The British Army began to distribute the helmets to troops in France in early May 1915. By early July 1915, all British troops on the Western Front had been issued with them. The French also used Hypo helmets to protect their troops from gas attack and, as Auld observed in Washington, the helmets 'stood up very well against chlorine'.[78]

Intelligence of a striking kind

In July 1915, 'we got word from our Intelligence Department of a striking kind', Auld reported in his address: 'It consisted of notes of some very secret lectures given in Germany to a number of the senior officers. These lectures detailed materials to be used, and one of them was phosgene, a gas which is very insidious and difficult to protect against. We had to hurry up to find protection against it.'

The chemists came up with the idea of using sodium phenate in the helmets. The chemical, also known as sodium phenoxide or sodium phenolate, reacts with chlorine and chlorine-containing chemicals such as carbonyl chloride – phosgene. The phenate is very soluble in water and is easily prepared by the reaction of phenol, a coal tar chemical often referred to as carbolic acid, and sodium hydroxide, also known as caustic soda.

The British Phenate Helmet, or 'P helmet', was similar in design to the Hypo helmet. It consisted of layers of flannel dipped in a solution of caustic soda and phenol, which reacted together to form the phenate, and glycerine. Two circular glass eyepieces in tin rims replaced the rectangular window of the Hypo helmet. The P helmet also had an exhaust tube with a valve for exhaled air. For this reason, the helmet was sometimes known as the 'Tube helmet'. Like the Hypo helmet, the open end of the P helmet was tucked into the soldier's uniform.

The helmets were used by British troops during the first German phosgene attack on 19 December 1915. 'This attack was in many ways an entirely new departure and marked a new era in gas warfare', Auld remarked: 'We never had any actual evidence during the attack that phosgene was being used, as no samples were actually taken from the cloud'. Cylinders of the gas were captured later and Barley identified it as phosgene.[44]

The P helmet provided adequate protection against low concentrations of phosgene, but proved unsatisfactory against high concentrations. Furthermore, the two eyepieces afforded only limited visibility. Finally, the phenate impregnated in the flannel tended to leach out in rain or if the soldier sweated, causing blistering to the face and neck.

In view of these limitations, the British chemists searched for other materials that could absorb phosgene. Meanwhile, Russian and French chemists had shown that an aqueous solution of a commercially available, water-soluble white crystalline powder, known as urotropin, could destroy phosgene.[80] The full chemical name of urotropin, an organic compound, is hexamethylenetetramine, which is usually abbreviated to hexamine. A molecule of the compound contains four nitrogen atoms and six methylene groups, a methylene group being a carbon atom attached to two hydrogen atoms. The chemical is readily prepared by treating formaldehyde, a colourless liquid and one of the simplest of all organic compounds, with ammonia.

The British adopted the chemical for their P helmets, which then became known as 'PH helmets', the PH standing for 'phenate-hexamine'. The combination provided protection against phosgene at concentrations up to 1 in 1,000, whereas the ratio was 1 to 10,000 for phenate alone. Before use, the PH helmets were soaked in a solution of hexamine, caustic soda, phenol and glycerine. The British introduced the helmet in January 1916, not only for troops but also for horses.[81] Soon after, an improved version, the 'PHG helmet', was introduced, which had tight-fitting goggles attached to the mask.

As the war progressed, the Germans and French developed respirators and gas masks along similar lines to the British. In the early chlorine gas attacks, for example, German troops used pad masks dipped in sodium carbonate-sodium thiosulfate solutions to protect themselves. The French employed not only solutions of sodium phenate, but also alkaline solutions of other chemicals for their pads and gas masks, including the sodium salts of ricinoleic acid, a type of organic compound known as a fatty acid, and sulfanilic acid, an organic acid that contains both sulfur and nitrogen. The first, sodium ricinoleate, was used to protect against chlorine and tear gas, whereas the second, sodium sulfanilate, protected against phosgene. With the threat of hydrogen cyanide gas attacks looming, they also used pads containing nickel acetate, a chemical that counteracted the gas.

The so-called French 'P2' masks employed three pads, one impregnated with a sodium ricinoleate solution, another with a sodium sulfanilate solution, and one with the nickel acetate solution. Their later 'M2' fabric mask contained a thick pad that covered the mouth and nose, and was prepared from around thirty layers of muslin dipped in a variety of solutions. The French manufactured over 29 million M2 masks between February 1916 and the end of the war. During the second half of 1916, millions of PH helmets and M2 masks were being manufactured by the British and French, while hundreds of thousands of these were shipped to the Russians.[82]

Use of canisters

An early and major development in gas mask design involved the use of filter canisters containing chemical materials that absorbed or destroyed the toxic gases. By the end of 1915, the Germans had already introduced a snout-type canister mask with a socket below the eyepieces. The canister contained absorbent materials and neutralising chemicals, and was screwed into the socket of the mask. The soldiers breathed in and out through the snout canisters, and when the chemicals in a filter were exhausted the canister was replaced with a fresh one.

Each canister contained three layers: the inside layer consisted of pumice, a volcanic rock containing lots of small cavities, mixed with hexamine to counteract phosgene; an outside layer, coated with fine-powdered charcoal, contained kieselguhr, a naturally-occurring porous solid, soaked in potassium carbonate solution to neutralise chlorine; and a layer of charcoal granules was sandwiched between the pumice and kieselguhr layers.

The type of charcoal commonly used in gas masks to adsorb tear gases and toxic gases was an amorphous and highly porous form of the material known as 'activated charcoal'. In chemistry, adsorption with a 'd' and absorption with a 'b' are two different processes. Adsorption occurs when a substance adheres to the surface of another substance, whereas absorption occurs when one substance passes through the surface of another substance and diffuses throughout the bulk of that substance. At the time of the First World War, both processes were commonly referred to as 'absorption' with a 'b', and activated charcoal was then classified as an absorbent.

Activated charcoal is prepared from the carbon obtained from wood, nut shells or other carbonaceous sources, and the activation process involves heating the carbon to high temperatures. Each gram of the activated material typically has an internal surface area of around 10,000sq ft. The material is therefore particularly useful for purifying gases and liquids as relatively large amounts of impurities can be adsorbed on the interior surfaces of the pores.

The power of activated charcoal as a toxic gas adsorbent was illustrated by a report published just after the war. Chemists at the Chemical Warfare Service in the United States revealed that activated charcoal was able to reduce a concentration of 7,000 parts of chloropicrin per million parts of air in a rapidly moving current of air to less than one part per million in less than a twentieth of a second.[83]

The British 'Large Box Respirator', also known as 'Harrison's Tower', was the earliest British respirator to employ activated charcoal in a canister. It was designed by British chemist Edward Harrison (1869–1918) and first deployed in February 1916. The respirator consisted of a facemask connected by a flexible corrugated rubber breathing tube to a separate filter canister that was carried in a haversack. The facemask was made of muslin soaked in a solution of hexamine and sodium zincate. The latter compound is an alkaline chemical formed by the reaction of sodium hydroxide with zinc or zinc oxide.

The canister typically contained granules of charcoal, potassium permanganate and soda lime. Potassium permanganate, a purple manganese-containing crystalline material, is a powerful oxidant that is able to react chemically with and destroy both toxic and tear gases. Soda lime is a whitish-grey granular mixture consisting of calcium hydroxide and a relatively small amount of sodium hydroxide. It may also contain a little potassium hydroxide. As soda lime is strongly alkaline, it counteracted acidic gases such as chlorine.

The Large Box Respirator was heavy and unwieldy, however. Furthermore, the facemask covered just the nose, mouth and chin, and although goggles were used separately to protect the eyes, they proved inadequate against tear gas. In April 1916, the British began to issue the 'Small Box Respirator' to troops. The respirator was like the Large Box version except that the face piece fitted tightly over the whole face and therefore protected the eyes against tear gas.

Other Allied countries either developed and produced their own gas masks with British-style designs or they used the British respirators. The Italians, for example, adopted the British Small Box Respirators. Similarly, the Russians employed British Small Box Respirators, and they also produced the Zelinskii-Kummant mask, named after the chemist N.D. Zelinskii and engineer E.L. Kummant who together designed and developed it.[84] Zelinskii was director of the Central Laboratory of the Ministry of Finance and a former chemistry professor at Moscow University. The mask consisted of a tightly fitting rubber headpiece and a small canister containing just activated charcoal. It proved effective against many gases, but unfortunately the mask was uncomfortable to wear and the charcoal tended to escape from the canister.

The French developed and produced not only the M2 Mask but also a range of other masks. Their 'Tissot Mask', for example, comprised a rubber facemask connected by tube to a canister of gas-removing materials, which was carried on the soldier's back. The mask overcame one of the major problems of other types of respirator, namely misting or dimming of the eyepieces caused by condensation, by drawing in air across the eyepieces. The French subsequently developed the 'ARS (*Appareil Respiratoire Spécial*) Mask' that incorporated a snout canister and the non-dimming features of the Tissot Mask. Over five million ARS Masks were manufactured between November 1917 and the end of the war.

The Germans essentially retained their snout canister design for gas masks until the end of the war. They made only relatively minor modifications apart from changing the contents of the canister as new war gases were introduced by the Allies. When the United States declared war on Germany in April 1917, 'almost no information had come to this country [USA] about gas warfare or the requirements of gas defense equipment', comments Bradley Dewey, a colonel in the Chemical Warfare Service, in an article published in 1919.[85] At the time, the

Allies had imposed a total news blackout on gas warfare[86] and, as Dewey explains, surrounded the whole subject with considerable secrecy.

Yet by the end of the war, the United States had manufactured over 5.25 million gas masks, which were box respirators modelled on the British Small Box Respirator design. The standard canister used with these respirators contained an 'intimate mixture of 60 per cent charcoal and 40 per cent "purple" soda lime, by volume'.[87] Purple soda lime was essentially a mixture of soda lime and sodium permanganate, a powerful oxidant like its cousin potassium permanganate. The mixture, which was widely employed in various types of American respirators, became known as the 'war gas mixture'.[83] Respirators were employed not just to protect troops from gas, but also to protect horses, mules, and message-carrying dogs and pigeons. A typical horse respirator, for example, consisted of a bag made of several layers of cloth impregnated with chemicals such as potassium carbonate and hexamine that counteracted the poison gases. As a horse does not breathe through its mouth, the bag was just fitted over its nose and upper jaw, and a pad was inserted into the horse's mouth to stop it chewing through the mask. Eyes were left unprotected as tear gas was not considered to seriously affect horses.

Other forms of protection

Gas masks and respirators were not the only form of individual gas protection employed in the First World War. As previously mentioned, protection against chemical agents such as mustard gas required full protective clothing; however, the protective overalls that were available tended to be heavy and especially uncomfortable on hot days when the vesicant liquid readily vaporised and the troops inside the suits sweated. By the end of the war, the Americans were beginning to develop improved full protective clothing against mustard gas. One such improved suit was a 'porous fighting suit, which allowed air and moisture ingress and egress, but which absorbed mustard vapour'.[64]

Gas-proof shelters also offered 'collective protection', as Prentiss puts it, against gas attacks.[88] They were either gas-tight, non-ventilated enclosed spaces or ventilated installations with some means of drawing in fresh air from outside while also removing toxic chemical agents from the air in the same way as a gas mask canister. Other gas defence measures included the provision of specially woven all-cotton blankets treated with oil to be used at the entrances of dugouts to make them gas proof.[85] Trench fans consisting of canvas flappers on wooden handles were also employed for removing gas from trenches and dugouts. Finally, ointments applied to the skin provided some protection against mustard gas burns. The ointments were typically mixtures of zinc oxide, lard, linseed oil and lanolin – a yellow waxy material obtained from sheep's wool.

Decontamination was a key weapon in the armoury of gas defence. After the first chlorine attacks took place, the British issued hand-operated agricultural crop sprayers, known as 'Vermorel' sprayers, filled with sodium thiosulfate solution to clear chlorine gas that had sunk into trenches, dugouts and shell holes. Before each spraying operation, troops assigned to the task reputedly joked: 'Let us spray!'[77]

Less persistent gases such as phosgene did not pose a problem: 'Nature is the greatest decontaminator' for such gases, Prentiss remarks, as they were readily dispersed by the wind. However, decontaminating areas affected by persistent agents, particularly mustard gas, was a real challenge, especially in enclosed areas or if the weather was cold and the oily liquid did not evaporate so readily. The operation had to be carried out by troops in full protective clothing using sand, ashes, soot or sawdust spread over the affected areas to absorb the liquid and prevent it from vaporising. However, these materials do not chemically react with and destroy mustard gas. Sand was sometimes mixed with a chemical such as calcium hypochlorite, the main constituent of bleaching powder. Although the powder reacted with mustard gas to form compounds that did not blister, it did so explosively and was liable to drive off some of the mustard gas as a vapour. It was therefore necessary to use a mixture containing three times more sand than bleaching powder.

Water, especially hot water, proved useful as it reacted with and destroyed the vesicant, albeit slowly. And as water is less dense that mustard gas, it could be used to cover the oily liquid and prevent its vapour from escaping. Late in the war, the Americans employed a small number of mobile 'degassing' units for treating troops contaminated by mustard gas. The units included tank trucks that could hold 1,200 gallons of water. They were fitted with water heaters that provided hot water for portable showers. Mustard gas victims were withdrawn from the frontline and taken to these units where they were degassed: the troops thoroughly washed themselves with hot water and soap, and medical staff also flushed their eyes, ears, mouths and noses with sodium bicarbonate solution, before issuing them with fresh uniforms.

Garments contaminated with mustard gas were decontaminated by hanging them in the open air. If the weather was warm and sunny, it generally took two days of airing to clear the gas from the clothing. Alternatively, the clothing was steamed to remove any traces of the gas.

Gas! GAS! Quick, boys!

The exhortation 'Gas! GAS! Quick, boys!' in Owen's poem, *Dulce et Decorum Est*, might just as well have been directed at the chemists of the belligerent nations who attempted to tackle the challenges of chemical warfare. However, chemists were not used to the novelty and tempo of the work required for which they had volunteered, explains Ludwig Haber in his book on chemical warfare in the First World War.[89] The emphasis was on speed: 'In every country the changing nature of chemical warfare and the leap-frogging contest between offence and defence created particular difficulties for research and development ... Lack of time added to them'. Evans also comments in his book, *Gassed*, that chemical warfare in the First World War was a 'scientific struggle to gain the upper hand', and that the 'ingenuity and determination of scientists on both sides was taxed and tested'.[90]

On the offensive side, chemists raced to discover new chemical agents that were militarily more effective than those used by the enemy. The ease and cost of the manufacture of these chemicals, their toxicology, their persistency, the best methods of delivering and dispersing them – by shell or from cylinders for example – their ability to penetrate respirators and many other factors all had to be considered. On the defensive side, chemists raced to identify these new agents, to devise analytical tests for them, and to develop respirators and other means of protection against them.

According to Ludwig Haber, more than 6,000 scientists in the belligerent nations were engaged in chemical warfare research development by the end of the war.[91] In Germany, the Kaiser Wilhelm Institute for Physical Chemistry and Electrochemistry, located in Dahlem, Berlin, was put under military control when the war broke out. Under the direction of Ludwig's father, Fritz Haber, the institute became the country's central research laboratory for the development of chemical weapons.

The institute was founded in 1911 and its first research building opened in October 1912. In 1913 and 1914 the institute had a staff of just five scientists, ten assistants, and thirteen volunteers and students. By the end of the war, however, it had a staff of over 1,000. In 1953, the institute was incorporated into the Max Planck Society for the Advancement of Science, an association of some eighty German research institutes, and renamed the Fritz Haber Institute of the Max Planck Society.

Following the first German gas attack in April 1915, the French developed a programme of chemical warfare research that was carried out in thirteen civilian and military laboratories in Paris.[92] The programme was divided into two sections, one on aggressive warfare and the other on protection. Some twenty eminent academic chemists and pharmacists, including Grignard, and their assistants were engaged in the programme. Their work was co-ordinated by the *Inspection des Études et Expériences Chimiques*, the French research and development organisation for chemical warfare.

In July 1915, the Russian War Ministry established the Commission for the Production of Asphyxiating Gases (CPAG). The CPAG had responsibility for the study, development and factory production of chemical warfare agents, for training troops in their use, and for organising special army gas units.[93] The Chemical Section of the Army Sanitary and Evacuation Service had responsibility for research on and the production of gas masks. Non-governmental groups in

Russia, such as the one led by Zelinskii, also developed gas masks. In February 1916, the Russian military created the War Chemical Committee to oversee gas warfare, including the production of gas masks and the training of troops. The committee soon placed orders for millions of the Zelinskii-Kummant masks for issue to the Russian Army.

In Britain, research on the offensive aspects of chemical warfare was carried out in chemistry departments at Imperial College in London, Birmingham University and other universities around the country.[94] The research was funded by the Ministry of Munitions, which had been established in June 1915, through its Trench Warfare Department. In a book on chemical and biological warfare, British authors Robert Harris and Jeremy Paxman point out that thirty-three different British laboratories examined some 150,000 known organic and inorganic compounds as potential chemical warfare agents and that virtually every leading chemist in the country was engaged in some aspect of chemical warfare work during the war.[95]

The Royal Army Medical College conducted defensive chemical warfare research at its laboratory in London's Millbank for the War Office's Anti-Gas Department. In the summer of 1915, the BEF also opened a Central Laboratory for chemical warfare at St Omer, France. The laboratory carried out chemical analysis work to identify the chemical warfare agents used by the Germans. However, it was not until October 1917 that Britain brought offensive and defensive chemical warfare research together under one umbrella. The Ministry of Munitions then established the Chemical Warfare Committee which took responsibility not only for all aspects of research on chemical warfare equipment and materials, but also for their development, production, supply and use.

In addition to research on chemical warfare at academic institutions, the Ministry of Munitions set up an experimental station at Porton Down in Wiltshire to study chemical warfare scientifically. The site, then known as 'The War Department Experimental Ground', opened in March 1916 with the construction of two army huts: one was used as an office and the other as a store, with the first trials beginning two months later. In June, Arthur Crossley (1869–1927), a professor of organic chemistry at King's College, London, was appointed head of the establishment.

Later in 1916, an experimental company from the Royal Engineers, assisted by a Royal Artillery Experiment Battery and officers of the RAMC, began to carry out testing at the site. Chemical, anti-gas and physiological laboratories were also established at Porton, as well as animal houses, a field trials department and a meteorological section.

Experiments initially focused on the dissemination of gas from cylinders, before the Porton station then switched to trials on the use of artillery shells, mortar bombs and grenades. The station also performed experiments on the persistency of poison gases and examined the toxicity of some 200 potential chemical warfare agents.[96] In May 1917, the station began work on the development of respirators, which involved testing their ability to prevent the penetration of new chemical agents such as arsenical smokes.

By the end of the war, the establishment had grown from its original two huts on 3,000 acres of land to a large hutted camp on 6,000 acres. In November 1918, the staff at Porton totalled some 50 officers and 1,100 other ranks, as well as 500 civilian staff.

The United States surges ahead

The Americans knew little of chemical warfare when they entered the war but, as Ludwig Haber remarks, 'once they got into their stride, outdistanced everyone else'.[97] Initially, the United States Bureau of Mines offered its research facilities and services to the country's War Department, which accepted and the bureau assumed responsibility for chemical warfare research. Mining problems in the country had already led the bureau to carry out exhaustive studies of poisonous gases and the use of gas masks for protection. The bureau allocated various chemical warfare research projects to industrial and government laboratories as well as to laboratories at American universities such as Johns Hopkins, Princeton, Harvard, Yale and the Massachusetts Institute of Technology.

The American University in Washington DC also offered its campus and buildings for war work. In July 1917, the bureau took over the university and transferred its chemical warfare research and development activities to the American University Experiment Station, as it became known. From January 1918 until the end of the war, scientists-in-uniform in the Chemical Service Section of the US Army also conducted experiments on gases at a laboratory near Paris.[98]

Other agencies also carried out chemical warfare activities: for example, the US Army's Medical Department had responsibility for the development and production of gas masks and training individuals in their use; and the Ordnance Department had responsibility for the production and filling of gas shells. In September 1917, the AEF established the Gas Service to consolidate the chemical warfare operations of the separate agencies. The War Department in Washington DC also set up an Office of Gas Service within the department to centralise chemical warfare activities home and abroad, and to link with the AEF Gas Service. Finally, in June 1918, the US president, Woodrow Wilson (1856–1924), signed an order to bring all these chemical warfare activities together into a single agency within the War Department – the Chemical Warfare Service. Wilson also ordered the transfer of the Bureau of Mines research programme at the American University Experiment Station to the new service.

The service, headed by Major General William L. Sibert (1860–1935), took charge of the research, development, production and testing of toxic gases and gas shells, gas-defence equipment and materials, as well as the training of troops in offensive chemical warfare operations and in gas defence. By the end of the war, the service was producing chemical warfare agents at the rate of around 100 tons per day – double what Germany was able to produce, according to science historian Andrew Ede at the University of Alberta, Canada, in an article on chemical warfare.[99] Sibert also enlisted almost 1,300 scientists and engineers as officers in the Chemical Warfare Service, who received 'significant resources for research and development of both defensive and offensive equipment, tactics and training'.

One of the most notable if not notorious achievements of the service was the development and production of Lewisite – the 'dew of death'. Lewis, who developed the toxic chemical, became director of the Offensive Branch of the service's unit at the Catholic University in Washington DC. Other chemists in the service at the nearby American University Experiment Station also worked on the development of Lewisite. Manufacture of the toxic oily liquid began in October 1918, just four months after the establishment of the service.

Immediately after the war, various divisions of the Chemical Warfare Service published a series of articles in the American Chemical Society's monthly *Journal of Industrial and Engineering Chemistry*. The contributions describe specific aspects of the service's work on chemical warfare. In the February 1919 issue of the journal, for example, the service's Research Division chief, George Burrell, reports that 'gas masks were the first American-made equipment for the American soldier to arrive in France'.[64] He notes that different kinds of wood, coconut and other nut shells, lampblack, blood, seaweed and other carbon sources had been tested as sources of charcoal for the masks. Coconut shell charcoal proved particularly effective for the adsorption of a variety of toxic gases and was therefore extensively used in the canisters: 'A tremendous amount of experimentation has been done on the preparation of charcoal', Burrell continues. Most importantly, hydrocarbon residues in the charcoal pores had to be removed as they reduced the adsorptive power of the charcoal. The Chemical Warfare Service researchers showed that the charcoal could be cleaned up by treating it with steam at over 900°C for several hours.

Research Division scientists also carried out experiments on the composition of the soda-lime granules used in the gas mask canisters. The mixture they finally settled on and used in their masks contained lime, cement, kieselguhr, caustic soda, sodium permanganate and water. Each ingredient performed a specific function: the lime, which is a chemical also known as calcium oxide, neutralised acidic gases; the cement hardened the granules without diminishing the porosity that was essential for good adsorption of the toxic gases; and the kiesulguhr increased the porosity of the granules. Burrell continues: 'The caustic soda activates the rate of absorption of most gases, and, in addition, makes the control of the drying process more simple and tends to maintain the proper water content in the finished granule'. The sodium permanganate was 'used primarily to oxidize certain oxidisable gases' which were difficult to adsorb in gas masks. Finally, the water was necessary to make up the mixture and 'get the best results in the absorption of gases'.

Condensation of droplets of moisture on the inner surfaces of gas mask eyepieces posed a serious problem for the American masks as it fogged the inner surfaces of the eyepieces and consequently impaired vision. Researchers at the American University Experiment Station therefore searched for an anti-dimming composition that would overcome the problem. After testing various experimental preparations, including soaps dissolved in volatile solvents, they settled on soapy mixtures containing Turkey red oil, sodium carbonate or sodium hydroxide, water glass and paraffin oil.[100] Some Turkey red oil preparations also contained glycerine. Turkey red oil, also known as 'sulfated castor oil', is made by treating castor oil, or other vegetable oils, with sulfuric acid. Unlike castor oil, which does not dissolve in water, Turkey red oil is soluble in water and can therefore be applied to surfaces as an aqueous dispersion. Water glass is a syrupy concentrated aqueous solution of sodium silicate.

The Turkey red oil anti-dimming mixture was dried and produced either as a paste or as a small stick. The instructions for use were: 'Wipe the surfaces of the eyepieces of the mask clean. Breathe on the inside surfaces and apply a little of the composition, rub it in thoroughly with the finger, then polish gently with the rag'.

7

Dye or Die

The boys in khaki

In Britain, nothing was more evocative of the First World War than the images of men in khaki uniforms marching through villages, towns and cities to cheering crowds, boarding troop trains and ships, marching along roads in France, fighting and dying at the front, or being tended to by khaki-clad medical officers in casualty clearing stations and hospitals.

The scenes and uniforms are depicted in numerous paintings, posters, postcards and cigarette cards. One example is the 1918 oil on canvas *Over the top* by John Nash (1893–1977), who was appointed official British war artist in April 1918. It shows a company of khaki-clad British troops setting out from their trenches to attack the Germans in December 1917. Another of Nash's 1918 paintings, *A French Highway*, portrays British soldiers in khaki uniforms marching along the 'Sacred Way' to Verdun, accompanied by two mounted French officers wearing dark blue cloaks.[1]

British Army recruitment posters aimed at women exploited the image of the 'boys in khaki'. They urged women to encourage their boyfriends, sons, brothers and husbands to enlist in the army and fight for 'King and Country'. The famous 'Women of Britain say – Go!' poster of 1915, for example, shows two women and a child gazing out of window at four soldiers in khaki uniforms as they march away. Another British poster, addressed 'to the young women of London' asks: 'Is your "best boy" wearing khaki? If not, don't you think he should be?'

Reports from the front frequently allude to khaki. For instance, a *British Medical Journal* special correspondent based at a hospital in a coastal town of northern France writes in 1914 that 'there are so many khaki-clad individuals of one kind and another to be seen around the quays that one might almost imagine that these rather than the black-coated passers-by were the normal inhabitants of the town'.[2]

Khaki is a drab yellowish or greenish brown colour. It is also the name of cotton or woollen fabrics or garments – especially military uniforms. The word derives from the Hindustani word *khak* which means dust, earth or ashes. As far as the British were concerned, khaki was the colour of the First World War, although it had not always been so.

The decline of brightly coloured battledress

The widespread adoption of standardised military uniforms did not become the norm for European armies until the seventeenth and eighteenth centuries. These early uniforms were mostly made of fabrics dyed with bright colours, notably red, blue, green and yellow. Before the advent of synthetic dyes in the second half of the nineteenth century, textiles used for the uniforms were coloured by natural dyes and pigments. Dyes are soluble chemicals that chemically bind to the materials they colour. Pigments, on the other hand, are insoluble natural or synthetic chemicals that do not chemically bind to the material they colour.

Brightly coloured military uniforms served a variety of functions, of which visibility on the battlefield was perhaps the most important. Before the advent of smokeless powders, firing weapons with gunpowder generated clouds of black smoke, so highly visible uniforms allowed soldiers on one side to be easily distinguished from enemy soldiers in the so-called 'fog of war'.

The introduction of breech-loading rifles in the nineteenth century sparked a gradual transformation in battledress. Unlike the muzzle-loaded, smooth bore muskets that had preceded them, rifles could fire accurately at long range. When exposed in the battlefield, troops in bright uniforms proved easy targets for enemy sharpshooters.

As the nineteenth century progressed, armies began to realise that drab-coloured uniforms provided camouflage in the battlefield and, as a consequence, made it more difficult for the troops to be picked off as targets. The use of bright colours began to be phased out, although the process was slow: at the beginning of the First World War, troops in some armies, notably the French Army, were still wearing brightly coloured uniforms in battle.

The British Army first used khaki for military uniforms in the late 1840s during a series of battles and wars against Afghan tribesmen, the Pashtuns, along the North West Frontier, which lies in a mountainous area that now separates Pakistan and Afghanistan. The khaki uniforms were introduced by Harry Lumsden, an officer of the East India Company's forces. He observed that the shining white cotton uniforms of British and Indian troops were highly visible targets for snipers in the dry and dusty terrain of the North West Frontier. He therefore adopted the drab khaki field uniforms for troops in a regiment he had formed called the Corps of Guides.

Other examples of the use of drab khaki-type field uniforms by the British Army soon followed. Tunics dyed a brownish-grey-purple were worn by troops in the Scottish 74th Highlanders, who were fighting in South Africa against the forces of the indigenous Xhosa people in a war known as the 8th Xhosa War (1850–1853).[3] This was the first time that khaki-type uniforms were used outside the Indian sub-continent. In 1868, a British expeditionary force of Indian troops was sent to Ethiopia, then known as Abyssinia, to secure the release of European captives that the Emperor of Abyssinia was holding. The troops wore white summer uniforms that had been dyed a greyish shade of khaki.

The British Army subsequently adopted khaki uniform as regulation dress for British troops in foreign service, but not for those at home. An army led by Major General Sir Herbert Kitchener was largely dressed in khaki when it reconquered the Sudan in 1898. In the same year, soldiers in the United States Army wore khaki uniforms for the first time in the Spanish-American War, a war in which the United States gained the remnants of the Spanish empire: Cuba, the Philippines and Puerto Rico. Khaki soon found other uses, for example, and around 1900 the United States government began to employ khaki-coloured fabrics for making not only uniforms, but also tents and kits. The US army officially adopted khaki uniforms in 1902.

British soldiers wore khaki field uniforms during the Second Boer War (1899–1902) – one of the wars that were fought between the British and Boers (also known as Afrikaners) for the conquest of southern Africa. The khaki uniforms made the British troops less conspicuous targets than they had been in First Boer War in 1881 when the troops in the British infantry wore scarlet coats: 'The peculiar shade of the terrain of South Africa made it possible to conceal the presence of troops from the enemy by adopting a shade for uniforms which blended with the colour of the landscape', observed John C. Hebden (1862–1929) in an address in 1918 to the New York Section of the London-based Society of Chemical Industry.[4] The British soldiers later found that the khaki cotton cloth used for their uniforms not only wore out quickly, but also did not keep them warm enough. They were therefore issued with khaki uniforms made of woollen serge, a more durable and warmer fabric. In 1902, khaki serge uniforms were adopted by the British Army as soldier's dress at home as well as abroad, except on full dress ceremonial occasions.

By the beginning of the First World War, most armies wore drab-coloured field uniforms. For instance, the British, Canadian and Russian armies wore khaki field uniforms, and the Austrian-Hungarian, German and Italian armies wore blue-grey, field grey and greyish-green uniforms respectively. There were exceptions, however, as French troops entered the war wearing blue coats and bright red trousers. Such uniforms proved a distinct disadvantage in the initial phase of the war on the Western Front which was characterised as a war of movement. As the French infantry advanced against the Germans over open country, they proved highly visible and were easily gunned down. In August 1914, for example, five French armies carried out an offensive in Alsace-Lorraine, which had been held by Germany since 1870, and the French troops suffered around 250,000 casualties in just four weeks. During the Battle of the Frontiers, which took place between 14–24 August along the eastern frontier between France, Germany and southern Belgium, 27,000 French soldiers died in one day alone – 22 August. The war of movement on the Western Front ended the following month when the Germans dug themselves in during the First Battle of the Aisne on 15–18 September. However, even with the emergence of trench warfare, French troops continued to wear brightly coloured uniforms.

In the first chapter of his book *Storm of Steel*, Ernst Jünger, who served in the German Army on the Western Front throughout the war, refers to the coloured uniforms when he describes his experiences in the chalk trenches during the First Battle of Champagne.[5] The battle, which was fought with long pauses between 20 December 1914 and 17 March 1915, resulted in around 90,000 French casualties. Jünger writes of '… vast sugar-beet fields, where we could see the luminous red trousers of dead French attackers dotted about'. The French armies were issued with less conspicuous bluish-grey field uniforms later in 1915.

In his address in New York in 1918, Hebden notes that 'khaki-dyed fabrics are used almost wholly for military purposes'. When the United States entered the war in April 1917, the country planned for an army of one million men who would require an estimated 30–40 million yards of khaki-coloured cotton and woollen fabrics for their shirts, tunics and overcoats.

Natural and synthetic khaki dyes

A variety of natural and synthetic materials have been used to dye fabrics and uniforms a khaki colour. In the mid-1800s, the British and Indian troops fighting along the North West Frontier in the Indian sub-continent used natural materials such as mud, coffee, tea leaves and dung as dyes to camouflage their uniforms. However, the colour produced by these materials was easily washed out and also faded in daylight. In other words, the dyes were not fast or permanent. The colours of fast dyes do not change when the dyes or dyed fabrics are exposed for long periods to light, steam, high temperatures or other environmental factors.

Another natural material, a brown dye known as 'cutch', was also employed to produce the khaki fabrics required for military uniforms in the nineteenth century. Cutch is extracted from the wood of acacia trees, also known as myrtle trees, by boiling the wood with water. The dye consists of a mixture of chemicals including a relatively high concentration of a family of organic compounds known as 'tannins'. Tannins are used to convert animal hides and skins into leather in a process known as 'tanning'. Cutch was also widely used for printing 'calico' – a plain white cotton cloth – in India's cotton fabric industry.

A variety of processes for producing khaki fabrics were developed to cater for the increasing demands of the military during the First World War. In his 1918 address in New York, Hebden outlined several methods of dyeing cotton or wool with the colour. Some processes involved the use of naturally-occurring or synthetic coloured organic dyes, while other methods employed dyestuffs, also known as 'mordant' dyes.

Mordant dyes are organic dyes that require a mordant to be effective. Mordants are typically mineral pigments such as insoluble metallic salts or inorganic oxygen-containing compounds known as oxides, that are applied to fabrics in order to stabilise the organic dyes. The mordant combines with the dye to form an insoluble product in the textile fibres, improving the fastness and quality of the dyed fabric. Cochineal, a red organic dye extracted from cochineal insects, and the mordant tin chloride, an inorganic salt consisting of tin and chlorine, were employed to dye the cloths for British Army officers' scarlet tunics from the sixteenth century until 1952.[6]

One of the mineral pigments used for mordant khaki dyes was a naturally-occurring, chromium-containing inorganic compound known as chrome yellow. The chemical name of this yellow-orange solid is lead(II) chromate. The mordant was typically used with organic dyes such as a 'nitroaniline': a family of three solid nitrogen- and oxygen-containing organic compounds with brown, yellow or orange colours.

During the nineteenth and early twentieth centuries, synthetic organic dyes were mainly synthesised from organic compounds, known as intermediates, extracted from coal tar. Coal tar is one of the products of the pyrolysis, the destructive distillation, of coal. The tar is a complex mixture of hydrocarbons, phenols, nitrogen compounds and many other classes of compounds.

Nitroanilines were synthesised from aniline, one of the dye intermediates extracted from coal tar. The intermediate was used to manufacture a range of synthetic organic dyes including 'mauveine', the purple dye discovered by English chemist William Henry Perkin in 1856. Twenty years after Perkin's discovery, French-born chemist Charles Dreyfus (1848–1935) founded Clayton Aniline, a Manchester-based company that manufactured dyestuffs. During the First World War, the company produced some of the khaki and blue dyes for the British Army and Royal Navy, and also won a contract for producing 1,500 tons of TNT for the war effort.

One of the leading researchers at Clayton Aniline before the war was Dreyfus' nephew, French-born chemist René Lévy (1875–1912). Lévy sailed on the *Titanic* on 14 April 1912 and, as the ship sank, gave up his seat on one of the ship's lifeboats to a fellow female passenger. He bid farewell, stayed on deck and was never seen again. Lévy was posthumously awarded with a special Royal Society of Chemistry President's Award for his 'outstanding act of gallantry' in 2012.[7]

Before the First World War, the United States imported the synthetic organic dyes it needed for its military uniforms from Germany. They included vat dyes, which are oxygen-containing, water-insoluble dyes that are unable to dye textile fibres directly but which are treated with a solution of a chemical, typically sodium hydrosulfite, in vessels or tanks known as vats. The process produces water-soluble colourless derivatives, called 'leuco compounds', which fabrics absorb when immersed in the vat. When the textiles are exposed to air, the leuco compounds are converted back to the water-insoluble dye and the colour of the dye is restored. Indigo, a blue dye used for making denim, is an example of a vat dye. In the First World War, indanthrene yellow, a brownish-yellow organic compound, was employed as a vat dye for colouring uniform fabrics. By mixing it with other dyes, it could be shaded to produce the khaki and olive-drab colours that blended with the terrain in Europe.

Vat dyes are not only fast to light and water, but also to acids and bleaches. With the onset of the war, the supply of these dyes from Germany dried up and the United States' military had to switch to the use of water-insoluble dyes containing sulfur. Sulfur black and sulfur brown dyes were typically produced by the reaction of an organic compound, for example a nitroaniline, with the chemical element sulfur and the inorganic salt sodium sulfide. Sulfur dyes were less expensive than vat dyes to manufacture, although their fastness to light was inferior. The military solved this problem by relaxing its standards for fastness.

In 1920, Grinnell Jones, chief chemist of the United States Tariff Commission, reported on progress of the American coal tar chemical industry during 1919, particularly in relation to the production of intermediates and dyes.[8] The chemist noted that there was a 'considerable decrease in the output of several intermediates required for making dyes used for army uniforms' after the end of the First World War. He pointed out, for example, that that there had been a 90 per cent decrease in the production of a nitroaniline required to make khaki dye for woollen uniforms.

Gatty's khaki

In 1884, a French inventor took out a patent for the discovery of a unique fast khaki dye that was to make an enormous fortune for his family.[9] The inventor, Frederick Gatty, discovered that khaki-coloured fabrics could be produced from cotton and woollen cloths or yarns dyed with a mixture of iron and chromium salts. As the dye consists of two minerals – the iron and chromium salts – it became known as 'mineral khaki dye'.

Gatty was born in Alsace, then a province of France, on 14 September 1819. Little is known about his early life except that on leaving school he was employed as a chemist in the calico printing industry. During this time, he invented an improvement for a process introduced by Frederick Steiner (1787–1869) in

England for colouring fabrics with Turkey red. This red pigment, also known as 'alizarin', occurs naturally in the roots of plants of the madder family. It was used as a dye in ancient Egypt, India and Persia, and was one of the natural pigments used to dye the fabrics required for the red coats worn by soldiers of various regiments of the British Army during the eighteenth and nineteenth centuries.

Gatty's improved dyeing process used 'garancin', a form of Turkey red dye extracted from madder roots using sulfuric acid. The dye was also known as 'Gatty red'. Steiner, who was also a native of Alsace, became aware of Gatty's discovery and invited him to come to England and join him at the Hagg Works at Church, a village near Accrington, Lancashire. Steiner had purchased the works in 1836 and used it for dyeing yarn with his Turkey red. Gatty travelled to England in 1842 and joined Steiner in a business partnership. The following year, Gatty established the company F.A. Gatty and Co. to dye yarn with Gatty red at the Hagg Works.

When Steiner died in 1869, Gatty took over the Hagg Works. Steiner's death more or less coincided with the first synthesis of alizarin from anthracene, a hydrocarbon obtained from coal tar. The synthetic alizarin was produced at a fraction of the cost of the natural Turkey red dye and, as a result, the use of the relatively expensive madder-derived red dyes fell into serious decline. Gatty's company continued dyeing yarn with the more expensive natural dye but also diversified into calico printing.

In 1882, Gatty visited India for a meeting with the Indian agent who marketed his company's Turkey red dyed cloth. During the trip, Gatty travelled to Shimla, then called Simla, a hill resort in the north-west Himalayas and the summer capital of the British Raj in India. He met an Indian Army officer who was on leave at the resort. The officer informed Gatty that British and Indian troops in some regiments were suffering high levels of casualties because of the visibility of their white hot-weather uniforms. To stop themselves becoming easy targets, the troops were camouflaging their uniforms by dyeing them with a tawny-colour which they called khaki. The Indian officer told Gatty that the khaki colour of the uniforms was easily washed out and soon faded in daylight, and suggested that a fortune could be made by the first company that could produce fabrics coloured with a permanent khaki dye.

Gatty returned to England and set to work in his laboratory at the Haggs Works to find such a dye. After numerous experiments he finally patented his process for producing a fast khaki dye on 20 August 1884. Gatty sent samples of military service uniforms dyed with the dye to India where they were tried out by troops on active service in the North West Frontier. They confirmed that the khaki-colour did not fade in sunlight and could not be washed out with water.

F.A. Gatty and Co. soon adapted their Hagg Works for khaki dyeing.[9] The process involved saturating the cloth or yarn with a solution of the iron-containing chemical ferrous acetate and the chromium-containing compound chromic acetate. The cloth or yarn was then developed with a solution of sodium carbonate and sodium hydroxide. The procedure resulted in the fibres becoming impregnated with a khaki-coloured mixture of iron- and chromium-hydroxide pigments and related chemicals.[10]

Gatty's mineral khaki dye revolutionised the uniforms of British troops in service in the tropical areas of the British Empire.

The capacity of the Hagg Works soon proved insufficient to meet the increasing number of orders for khaki-dyed fabrics required for the uniforms of British and Indian troops. In 1896, the company relocated its khaki-dyeing operations to a larger site at the Bannister Hall Print Works in Walton-le-Dale, a town near Preston, Lancashire. The buildings were modified and new machinery installed, and the business continued there under the management of the Gatty family until 1930.

The family made a considerable fortune from its khaki-dyeing process, although Frederick Gatty himself was not to enjoy the full fruits of his discovery. He died on 24 June 1888 and one of his four sons, Victor Herbert Gatty, became managing director of the company.

The Gatty family lived in Elmfield Hall, Accrington. During the First World War, the family placed the hall at the disposal of the British War Office. The hall was fitted out as an auxiliary army hospital and wholly maintained by the family. After the war, the family donated the hall and its grounds to Church Urban District Council. The park, named Gatty Park, was opened on 26 June 1920 for the purpose of preserving an open space for the recreation of local people, and has a memorial to the men of Church who died in the First World War.

When Victor Gatty died in 1922, at the age of 58, his brother Frederick Alfred Gatty, who was Gatty senior's eldest son and a chemist, mathematician and gifted musician, took over the management of the company.

8

Caring for the Wounded

The wounded infantryman

A shard of metal from the shell burst pierced the thigh of the infantryman. He collapsed into the bottom of the trench where he lay dazed with concussion and bleeding profusely. A doctor and an ambulance man from the RAMC crept along the trench towards him, cut away his trouser leg and applied an antiseptic dressing and bandage. The wounded soldier recovered consciousness but was unable to leave the trench because of the continued enemy shelling. He lay there damp, hungry, cold, in pain, in shock and losing blood. When night fell, the shelling abated and allowed stretcher bearers to carry him away from the trench through the dark woods. They took him first to the regimental aid post and then to the dressing station, which was an improvised hospital in a village church with a floor covered with straw.

There, overworked and exhausted doctors tended to the increasing numbers of wounded brought in. The infantryman was covered with a blanket and left to go to sleep. The following day, the infantryman had his dressing replaced. Then, although he was feeling feverish and ill, he was moved out to make space for more wounded who continued to arrive at the dressing station.

An ambulance took him and other wounded along a rough road to an ambulance train consisting of horse-boxes carpeted with straw and some third-class carriages. It was not possible to attend to the dressings of the wounded in the horse-boxes. The train eventually departed, moving slowly and jolting. After frequent delays, the train arrived at Rouen two days later. Motor ambulances collected the sick and wounded from the train. The infantryman and his companions were eventually taken to Le Havre and then across the English Channel in a hospital ship to Southampton, before they were finally transferred to a hospital in Britain.

This 'journey of a wounded soldier' was typical, according to the medical correspondent of *The Times* who told the story in a report from Paris on 17 October 1914.[1] As was usual during the war, such news reports were low on specifics and the correspondent did not mention the names of the village or the infantryman himself. However, the story alluded to the battle front running from the River Aisne region south towards Paris. The wounded soldier was therefore almost certainly a professional soldier in the BEF that was sent to France in August 1914. In the First Battle of the Marne, 5–9 September 1914, the BEF and French armies halted the German forces that had crossed the River Marne en route to Paris. A few days later, in the Battle of the Aisne, 15–18 September, the Allies pushed the Germans back across the River Aisne: trench warfare on the Western Front had begun.

Nine stages

The transfer of sick and wounded soldiers from the trenches to hospital was a complex, difficult and dangerous operation. British soldiers who were wounded or fell sick on the Western Front were transported along routes known as 'lines of communication' to medical bases at Channel ports in France and Belgium, and then to hospitals spread around Britain. Conversely, the sick and wounded of the ill-fated Gallipoli campaign, which took place from April 1915 to January 1916, were transferred to base hospitals on the Greek island of Lemnos, Alexandria or Cairo in Egypt, or on the island of Malta.

A report in the 14 November 1914 issue of the *British Medical Journal* described nine stages in the journey of sick and wounded soldiers from the frontline to the home hospitals.[2] Triage was carried out at several stages. Priority medical treatment was often given to those who had a good chance of recovery and could be sent back to the front or returned home. Soldiers who had little chance of survival were, where possible, made comfortable, provided with nursing care, given morphine and then left to die.

In some lines of communication, two or more of these stages coalesced into one. For example, in some instances the field ambulance was also the casualty clearing station, and in the Gallipoli campaign hospital ships acted as casualty clearing stations as well as first-line hospitals.

Stage 1: First field dressings

Every British soldier carried two small first aid packets, each containing a pad of antiseptic gauze and a bandage to dress entry and exit wounds measuring no more than 3–4in. German soldiers were also supplied with similar field dressings, while French troops used aseptic rather than antiseptic gauzes.

The dressings were applied by the wounded soldier himself, or by a comrade. After dressing the wound, the injured soldier could leave the firing line if the military situation allowed. Regimental and field ambulance stretcher bearers carried more heavily wounded soldiers to the regimental aid post. The dressings were often left undisturbed until the soldiers reached the casualty clearing station.

Stage 2: Regimental aid posts

The regimental aid post was staffed by the regimental medical officer (RMO), who typically carried his medical equipment on a two-wheeled cart, and a number of medical orderlies. The post was ideally located in a safe position near the firing line and a road leading to the field ambulance. At the post, first field dressings were sometimes replaced, splints and tourniquets applied where necessary, and morphine was given to those in severe pain. The RMO decided who should go to the field ambulance. The wounded then walked, if able, or were carried by stretch bearers or on horse-drawn wagons to the nearest field ambulance.

Stage 3: Field ambulances

The field ambulance was not a vehicle but rather a mobile medical unit located in tents, farm buildings, a village church or school, or some other place no more than a mile or so behind the frontline. The ambulance facilitated the transfer of sick and wounded men from several regimental aid posts to the nearest casualty clearing station. The work of the ambulance, however, was often hindered by atrocious conditions at the front: 'The mud everywhere, but especially between the regimental aid posts and the advanced dressing stations of the field ambulances, was amazing, and the labour involved in stretcher carrying enormous … It took practically six bearers to a stretcher [instead of the usual two]', notes a report from Ypres in August 1917.[3]

The ambulance consisted of advanced dressing stations as near the frontline as possible, stretcher bearer sections, a main dressing station some distance away, and other facilities such as a cookhouse, laundry and bathing station. The medical officer at the advanced dressing station made triage decisions about whom should be returned to the front, whom should be moved to the main dressing station for further treatment and triage, and whom should be transferred to a base hospital via a casualty clearing station. Because roads were often in a bad state, horse ambulances were sometimes employed to carry soldiers from the advanced dressing stations to the main dressing stations. If the roads weren't too damaged or muddy then motor ambulances were used.

Soldiers who were close to death were kept in the main dressing station. The ambulance staff tended to the sick, cleaned and treated injuries, provided wounded soldiers with hot drinks and food and, if required, more morphine. Only emergency operations were carried out at the ambulance, such as amputations where there was no other way of dealing with a limb.

The 14 November 1914 report in the *British Medical Journal* describes the ambulance as a collecting station for the wounded that offered little from a medical point of view: 'The ambulance – to put it bluntly – repacks the man for further transport, and takes his name etc., for the official returns'.[2]

Stage 4: Casualty clearing stations

In the next stage of the journey home from the Western Front, sick and wounded soldiers were transported from the field ambulances to casualty clearing stations, also known as clearing hospitals, by convoys of motor ambulances. Whereas it was impossible to give little more than first aid at dressing stations, casualty clearing stations could provide better treatment. However, even at these stations, 'after a big action, when large numbers of wounded pass through, it is only very few that can be dealt with adequately as the principles of wound treatment demand', comments RAMC medical officer G. Grey Turner in a report published in 1916.[4]

Casualty clearing stations were mobile units: typically makeshift hospitals with wards and operating rooms in tents, wooden huts, schools, monasteries, or other institutional buildings. They were generally located next to the sidings of a railhead in a town some 6–12 miles behind the frontline.

The journey from the frontline, via the regiment aid post and field ambulance, to the station often took between twelve and thirty-six hours. Medical and nursing staff at the stations replaced first-aid dressings, administered painkillers, took X-rays, carried out emergency operations and made more triage decisions. Some soldiers recovered sufficiently from their treatment at the stations to be returned to their battalions. Others, after spending perhaps a night or two at a station, were placed on ambulance trains or barges for transfer to a base hospital.

'At the front the most that can usually be done is to see that the wounds are well cleaned, freely opened and drained, hopelessly damaged tissues cut away, and fractures splinted in such fashion as to prevent movement while travelling', notes one account of the medical care of soldiers wounded at the Somme.[5]

A report published in March 1917 details the general procedure used at an advanced casualty clearing station for treating gunshot wounds of the abdomen.[6] Before being taken to the operating theatre, the patients were given a drug called scopolamine, also known as hyoscine, which prevents muscle spasm. They were also given omnopon, an analgesic that contains a mixture of salts of opium alkaloids such as morphine and codeine.

In theatre, the patients were anaesthetised with a combination of hot ether, an organic compound known as diethyl ether, and oxygen. Saline, a solution of sodium chloride in water, was given intravenously throughout the operations. The skin of each patient was prepared with a solution of picric acid, a pale yellow compound also known as 2, 4, 6-trinitrophenol, in 'spirit', that is picric acid dissolved in an aqueous solution of ethanol. The pale yellow solution was one of many types of antiseptics employed in the First World War to clean wounds and prevent infections.

During the operation at the advanced casualty clearing station, wounds were thoroughly wiped clean using gauzes wrung out of saline or antiseptic solutions. Following closure of the abdominal incision and recovery from the anaesthetic, the patient 'if restless' was given more omnopon. Any vomiting or hiccupping was relieved with 'small sips of brandy or champagne' or tincture of iodine, that is a solution of iodine in ethanol, diluted with water.

Stage 5: Ambulance trains and barges

Ambulance trains collected the sick and wounded at the railheads near the casualty clearing stations and transported them to a medical base. The average train consisted of about twenty coaches and was about 300yds long. The staff on each train included RAMC personnel and nursing sisters.

Canal ambulance barges, known as *ambulances flottantes*, were also used for this purpose. The typical barge journey, from Paris to Rouen for example, normally took one day including loading and unloading. During the Gallipoli campaign, wounded from the field ambulances on the peninsula were put on the barges which were then towed out to hospital ships.

Stage 6: Base hospitals

After a journey from the Western Front lasting many hours or even several days, the sick and wounded were finally delivered to a medical base, often established at a coastal town in Belgium or France, such as Étaples or Calais, or directly to a hospital ship.

Ambulance trains unloaded the patients at the base and medical officers selected an appropriate hospital for each patient. They sent some to general hospitals or 'convalescent camps', while others went to specialist hospitals, for example those specialising in fractures or infectious diseases. Some men were then sent home to Great Britain or Ireland, while others, after a short spell of convalescence, were returned to their units at the front.

The nature of the illnesses and wounds varied immensely. For example, one report published in February 1915 notes that servicemen arrived at the Anglo-American Hospital in Wimereux, a coastal town in France near Boulogne-sur-Mer, with bullet, shell, shrapnel and other types of wounds, and that the vast majority of wounds were 'in a more or less septic condition'.[7]

The report explains that medical staff at the hospital dressed lacerated wounds while the patients were anaesthetised with chloroform. The staff removed particles of hair, clothing, dirt, fragments of shell or shrapnel and bone, and irrigated the wound with hydrogen peroxide solution, an antiseptic and cleansing agent, or swabbed it with pure carbolic acid and then irrigated it with saline solution.

The base hospitals, unlike the mobile field ambulances and casualty clearing stations, provided 'the opportunity for finer kinds of surgery', as one medical officer at the Somme put it in July 1916.[5] A report in a September 1916 issue of the *British Medical Journal* outlined an example of the finer surgery and improved medical care at the base hospitals, referring to a surgeon and a bacteriologist working closely together in a large general hospital at an undisclosed base in France.[8] The bacteriologist used an antiseptic hypochlorite solution to reduce the number of microorganisms in wounds 'to a safe minimum'. The wounds were then sewn up resulting in 'a great saving of the time' which the patients would 'otherwise have had to continue under the surgeon's hands for daily dressings'.

Stage 7: Hospital ships

The ambulance trains transferred some of the sick and wounded from the Western Front directly onto hospital ships waiting at French ports. Motor ambulances carried sick and wounded from the base hospitals to the ships for passage across the English Channel to British ports such as Southampton.

Hospital ships also operated in the Mediterranean and elsewhere. In October 1915, Hubert Chitty, a Royal Navy surgeon, describes in vivid detail the work of surgeons on an unnamed hospital ship that carried wounded soldiers from the Gallipoli peninsular to Alexandria, Malta and ports in England.[9] The ship, which had two operating theatres, was able to carry 350 'cot cases' and the same number of 'walkers'. Its staff included eleven surgeons, four nursing sisters and St John Ambulance personnel. Most of the patients were suffering from shrapnel or bullet wounds, or injuries inflicted by bombs, and their survival depended on how long it took for them to reach the ship. Those who arrived early from the frontline were less likely to die from septic infection. The mortality rate among late arrivals was much higher, as these patients were normally exhausted and their wounds highly infected. Surgery in these cases normally consisted of incisions and amputations.

The journeys of these hospital ships were perilous as they often had to sail through mine- and submarine-infested waters. For instance, the hospital ship *Anglia* sank in the Dover Straits on 17 November 1915 after striking a mine laid by a German submarine. Over 130 soldiers, all of whom had been wounded, and a nurse were reported to have lost their lives.[10]

In another example, the hospital ship *Lanfranc* was torpedoed without warning on 17 April 1917 while carrying wounded to a British port. The secretary of the British Admiralty announced on 23 April 1917 that thirteen British wounded soldiers, one member of the medical staff, five crew and fifteen wounded

German prisoners were missing and presumed drowned. Over 150 wounded German prisoners were rescued by British patrol vessels 'at the imminent risk of being themselves torpedoed'.[11]

Stage 8: Ambulance trains

Specially equipped trains designed to carry sick and wounded troops from British ports to home hospitals made their initial runs in October 1914.[12] These trains were ambulances rather than hospitals and their interiors were designed to make the patient as comfortable as possible on the journey.

Movement in the so-called 'cot coaches' was minimised by hanging the cots in two tiers by chains from the carriage roof. A typical ambulance train had around 100 cots, including separate cots for officers, and accommodation for over 150 'sitting cases'. Sitting patients suffering from venereal diseases travelled in separate coaches. In one of the carriages, a dressing station stocked with drugs, surgical instruments and an operating table catered for emergencies, although, wherever possible, medical treatment and dressings were kept to a minimum.

The numbers of sick and wounded reaching home hospitals by ambulance train was astounding. A report in the *British Medical Journal* in 1915 reveals that between 19 September 1914, that is about six weeks after the BEF began to land in France, and 1 March 1915, one hospital alone, the Second Western General Hospital in Manchester, admitted 10,537 sick and wounded troops, an average of over 575 per week.[13] They were carried in seventy-three ambulance trains from Southampton, with each train carrying between 100 and 250 soldiers, and the journey took just over seven hours. The British Red Cross Society met each train and transferred the patients to the hospital in a fleet of ambulances and motor cars.

Stage 9: Home Hospitals

The final destinations for sick and wounded British soldiers on their journey along the lines of communication were the home hospitals – general, military, Red Cross and other types of hospitals in towns and cities in England, Ireland, Scotland and Wales. Their tortuous journeys from the Western Front to these hospitals often took a week or more, and by the time the patients had arrived they were often exhausted.

An exceptionally favourable dressing station

Each day during a three-week period along the frontline in France in September 1914, the stretcher bearers of one field ambulance conveyed between 20 and 300 sick and wounded soldiers from a mile-long section of trenches to a dressing station in a village called 'T', reports Gordon R. Ward, a captain in the RAMC Special Reserve, in his description of the station published in May 1915.[14] The station was located near a church on the side of a hill. To get there the bearers had to carry the sick and injured across fields. The number of medical officers at the station varied from three to five. Whenever possible regimental medical officers of the brigades in the nearby trenches also helped at the station. One hundred or so stretcher bearers of the field ambulance also occupied the station. Horse-drawn wagons bore the sick and wounded, after they had received treatment at the station, along shell-holed roads at night to a casualty clearing station some 5 miles away.

The field ambulance billeted over 100 of its staff and accommodated some of the sick and wounded in a farm in the village. The lodgings included a cattle shed where the wounded were laid on straw. A large barn also served as a ward for soldiers who were sick as opposed to wounded. Next to the barn was a small outhouse, which was used as a mortuary, and a large heap of manure. Another barn housed more wounded. Water streamed through the farm ending up in a filthy pool. The farmhouse itself was used to accommodate officers. The farmer lived in the farmhouse cellar and only emerged to feed his ferrets, according to Ward.

The German artillery soon found the range of the farm and shelled the buildings, hitting several men and wounding the farmer. The ambulance therefore evacuated the farmhouse and adjoining buildings and moved the sick and wounded to a series of caves in the limestone hillside at the top of a track behind the farm. The cave entrances were up a steep bank.

The wounded often arrived 'in the form of a rush towards nightfall', even though it was difficult to carry a stretcher in the dark along the track. The caves were fitted with mattresses found in a house in the village and the floors lined with straw.

The dressing station also suffered from a lack of skilled labour, such as nursing orderlies and sanitary men, and insufficient light, medical stores and comforts. Meals were cooked at the entrances of the caves as there was no ventilation for smoke inside, and the caves were infested with 'all manner of unpleasant insects'. The medical staff left good field dressings untouched and did not change the clothes of their patients even though they were dirty or blood-stained. In one of the caves, which contained the only good light, the staff dressed wounds, splinted fractured hips, and, on occasion, carried out amputations.

'The sanitation of the place was appalling,' Ward remarks, 'the whole hillside was one vast latrine'. Each fresh change of troops in the trenches above led to more latrines being dug. Furthermore, horses were stabled above one of the caves and a small stream ran down a path to the caves. The only disinfectant available was a little lime. 'The sanitary problem was almost insuperable … Yet this was an exceptionally favourable station' Ward concludes.

A tough chicken

Curiously, a dressing station in the caves featured in another story published in the *British Medical Journal* some eight months before the publication of Ward's description. The story, entitled 'Midwifery in the fire zone', was posted in an envelope, endorsed 'No military information' by an unnamed medical officer, possibly Ward, in the RAMC Special Reserve.[15]

The story describes a small village lying below limestone caves. Above the caves were 'reserve trenches, then trenches and wire entanglements, then the dead whom neither side could remove'. The caves housed the wounded and dressing station.

One day, the medical officer went with his servant to attend a woman who had just given birth to a baby in a cellar in the village, but there had been complications and the placenta had not appeared. Lysol and water were available, the RAMC officer reports: 'Thus the placenta was soon removed'. Lysol is a disinfectant containing organic chemical compounds known as cresols or 'hydroxytoluenes'.

The officer does not describe how the disinfectant was used. He does note, however, that the inhabitants of the village were pleased. They shouted '*Merci, Monsieur*', and offered him money, 'a practice not without merit but out of place in the circumstances'. An hour or two later, his servant arrived with a chicken donated by the villagers. It proved 'uncommonly tough', the officer observes.

The next day, the officer visited the mother with some medicines and discovered that she had no milk and was not able to produce any. 'Anyway, I gave her from His Majesty's stores two tins of condensed milk,' the officer reports, adding: 'I fear that the gift may have led to a violation of the principles of infant feeding as laid down by eminent authorities.'

The unnamed medical officer was relieved the following day and so was unable to see his patient again.

The recoveries

Once sick and wounded troops arrived at the home hospitals in Great Britain and Ireland, their chances of recovery were high. As early as 1915, 95 per cent of those who received treatment at these hospitals were being 'discharged to duty, on furlough, or sent to convalescent homes'.[16] About 1 per cent died and 4 per cent were regarded as 'permanently unfit'.

The percentage of recoveries near the firing line was much lower, however. The British campaign in Mesopotamia provided one example: between 5 December 1915 and 29 April 1916, the number of wounded British and Indian troops and civilians brought into hospitals at Kut, a city in Iraq, totalled 2,418, reports Major C.H. Barber, a surgeon in the Indian Medical Service.[17] The injuries were inflicted by shells and bombs falling on the city during the notorious Siege of Kut by the Turkish Army. According to Barber, 488 of the wounded died, 250 'remained for invaliding' and 1,680, almost 70 per cent, were returned to duty. Nevertheless, a good many of those returned to duty remained in poor health and were 'crippled in one way or another'.

Barber comments that surgical treatment and care of the wounded was hindered by the unfavourable conditions during the siege: the surroundings of the improvised hospitals were unhygienic; and there were insufficient personnel, equipment and stores. For example, the stock of antiseptic hydrogen peroxide solution was depleted early on in the siege. Inappropriate and insufficient food led to scurvy among the Indian troops and beri-beri among the British. Both diseases hindered the healing of their wounds.

The situation during the Siege of Kut may not have been typical, but reports from the Western Front indicate that the proportion of recoveries at casualty clearing stations was no better. For example, 500 cases of gunshot wounds to the abdomen were seen at one advanced casualty clearing station while active fighting was in progress.[6] Of these, 356 were operated on, 138 were moribund and six did not warrant surgery. Just 171, that is 48 per cent, of the 356 patients recovered from their abdominal operations.

Echlin S. Molyneux, a surgeon at another casualty clearing station, treated wounds of the abdomen during 'the final campaign' in 1918.[18] Of the thirty-four consecutive cases he saw during the campaign, twenty-three made a good recovery, but eleven died – a recovery rate of 67 per cent.

The recovery rates of soldiers injured in the First World War depended on a broad range of factors, not least of which was the military situation in which the wounds were inflicted. Speedy removal of the wounded from the battle scene and speedy treatment at a field ambulance and casualty clearing station were critical. The survival of the sick and wound also depended on the bravery, skill and endurance of the medical officers and orderlies, the stretcher bearers, the surgeons, and the nurses in the field ambulances, casualty clearing stations, ambulance trains, hospital ships and the hospitals at the medical bases and at home.

But these were not the only factors determining recovery rates. Successful medical treatment of the sick and wounded also relied on the adequate provision of hospital and medical equipment and, of course, a wide variety of chemical preparations including antiseptics, anaesthetics, disinfectants, analgesics and other pharmaceuticals – without these chemicals, recovery rates would have been much lower.

9

Fighting Infection

The wastage of war

Violence, disease, and death came in many forms during the First World War. German conscript Erich Maria Remarque succinctly illustrates this point in just a couple of lines in his novel, *All Quiet on the Western Front*:

> Shells, gas clouds and flotillas of tanks – crushing, devouring, death.
> Dysentery, influenza, typhus – choking, scalding, death.[1]

There are 'no other possibilities', he continues, than the trench, the hospital and finally the mass grave.

A book about violence in history, published in 2011 by Harvard University professor of psychology Stephen Pinker, includes a table ranking twenty-one wars and other violent episodes in history in terms of their death tolls.[2] The Second World War, with a death toll of 55 million, ranks top, while the First World War ranks number thirteen with a toll of 15 million deaths. The tolls include not only battlefield fatalities but also indirect deaths of civilians from starvation and disease. An estimated 6 million or so men in the forces of the belligerent nations were killed in battle during the First World War and another 7 million civilians died as a result of war. If we deduct these two death tolls from Pinker's toll of 15 million, we find that around 2 million men died from disease or from other causes such as accidents or suicide in the war.

War death tolls such as these are no more than broad estimates and often vary immensely depending on the source of the information. Similarly, records of the causes of death, if they were made at all, are generally imprecise and incomplete. Even so, it is possible to draw some conclusions from such records where they exist. In his handbook about the war, First World War expert Geoff Bridger provides statistics for the British Army showing that about 662,000 British soldiers lost their lives in the war. Around 438,000 of these soldiers were killed in action, 138,000 died of wounds and the remaining 86,000 died from non-military causes such as disease or accidents.[3] Such statistics do not show exactly how many men died of disease but they do suggest that no more than 13 per cent of the deaths at most could have resulted from disease, probably infectious illnesses caught on the battlefield, rather than from enemy action. Furthermore, approximately 21 per cent of the fatalities were not killed outright but died of battlefield wounds. Inevitably, a certain proportion of this 21 per cent died as a result of wound infections.

What is not in question is that the percentage of military deaths caused by disease during the First World War was markedly lower than in earlier wars such as the Crimean War and the South African wars of the previous century. For example, following Britain's entry into the Crimean War, some 20,000 British troops lost their lives. According to military historian Gordon Corrigan, 80 per cent of these deaths resulted from disease and just 20 per cent from Russian bullets.[4] Author Philip Hoare suggests the number of British deaths in the war was 17,000, but that the percentage of deaths from disease was 90 per cent: 'For every one of the 1,700 who died of their wounds, another nine would die of disease', he writes in his book about the Royal Victoria Hospital at Netley, near Southampton, which was built in 1856 to nurse wounded troops returning from the war.[5]

It was pathogenic, that is disease-producing, microorganisms rather than bullets that decimated the armies, remarked James Parkinson, president of the Medical Society of the State of California, in an address in 1910.[6] He was referring specifically to the thousands of healthy young American men who died from preventable diseases in reserve camps located in the United States at the time of America's intervention in the final year of the Cuban War of Independence (1895–1898) against Spain, the so-called Spanish-American War: 'The bacteria of pathogenic processes' conducted a twenty-four-hour campaign in these camps every day.

In the Second Boer War (1899–1902), of the 22,000 soldiers who died, disease accounted for 14,000, almost two thirds of the deaths, and bullets for just 8,000.[7] One of the British Army medical doctors who served in the war was Sir Alfred Keogh (1857–1936). During his service in South Africa, he soon realised that the high percentage of deaths and incapacitation from disease in the army, the 'wastage of war' as it became known, could be substantially reduced by taking a range of disease-preventing measures. In October 1914, when the British government appointed him director-general of the Army Medical Service, he was determined to organise efforts to improve the health and sanitation of the British Army and thereby minimise this wastage. The measures he adopted relied to a significant extent on a thorough scientific understanding of the nature and causes of infectious diseases in general and in particular on the use of chemical products, such as pharmaceuticals and disinfectants.

The squalor of war

Standards of cleanliness, hygiene, sanitation and care varied immensely depending on the location of the fighting and hospitals, availability of clean water, medical supplies and equipment, and training of the medical and nursing staff. Troops in the First World War frequently lived for days in rat- and fly-infested trenches and dugouts, where they were bitten by fleas and clothed in lice-ridden uniforms. They fought on muddy battlefields and sheltered from enemy fire in slimy shell holes and ditches. When wounded they were taken to makeshift first-aid posts, casualty clearing stations and hospitals, many of which were insanitary and overwhelmed with other sick and wounded troops.

Such problems sometimes started before a soldier even reached the frontline of battle. While waiting to board a hospital ship in 1914, C. Brehmer Heald, a Royal Naval surgeon, visited a number of isolated huts and tents occupied by troops of a battalion of British Territorials who were guarding the bridges and culverts along a 50-mile section of a railway line in England.[8] Heald discovered badly prepared latrines, some of which were too shallow, some swarming with flies and others too near tents. Many of the latrines were located in ditches leading directly to drinking or bathing water, while food and especially meat were exposed to flies, and dishes and utensils were not properly cleaned. These conditions inevitably lead to outbreaks of diarrhoea.

On the battlefield, wounds were often inflicted under filthy conditions and consequently became infected, as G. Grey Turner, commented in a lecture on military surgery delivered at the Royal Victoria Infirmary, Newcastle-upon-Tyne, in 1915.[9] Turner was a member of a team at the First Northern General Hospital in Newcastle who treated the wounded who had been shipped home from Flanders and the Dardanelles. Many wounded had to lie on the battlefield for hours if not longer: 'Men have spoken to me of lying for days' before receiving first aid, often having to lie in clothes that had been worn for days or weeks and in soil that had been richly cultivated and fertilised and was therefore teeming with microorganisms. Even at the casualty clearing stations some 8–10 miles behind the line, the volume of wounded was such that only a few could be treated adequately: 'The possibilities of sepsis are infinite'.

One such victim was a relation of the English writer Vera Brittain (1893–1970). In her autobiography she writes about a cousin from Ireland who received a minor bullet wound behind the ear at Suvla Bay during the August 1915 offensive of the Dardanelles campaign.[10] He lay unattended for a week before being operated on while 'suffering from cerebral sepsis' by an overworked surgeon on the hospital ship *Aquitania*. The cousin later died of his wounds.

Dead bodies on the battlefield also presented their own problems. In *Goodbye to All That*, war poet Robert Graves refers to a canal at Cuinchy, a village in northern France, that bred rats: 'They came up from the canal, fed on plentiful corpses, and multiplied exceedingly'.[11] Frank Richards, a private in the British Army, was also troubled by dead bodies. In his account of the war he describes a ditch that ran from the enemy's trench through no-man's-land to his own trench at Bois-Grenier – another village in northern France.[12] Richards and his fellow soldiers drew water from the ditch 'for drinking and cooking purposes', and one night, while on a patrol, they discovered dead bodies lying in it. Even so, they still continued to drink the water.

Conditions in some hospitals were similarly dire. A report in the 5 June 1915 issue of the *British Medical Journal*, for example, provides a graphic description of the wards in hospitals at Niš and Kragujevac, two cities in Serbia.[13] For a time, the hospitals lacked sanitary arrangements, baths and even beds, and men died like flies every day from 'sheer neglect'. The fever hospital at Kragujevac was shockingly dirty and alive with vermin; there was virtually no nursing and as the patients were not even washed, they because filthier and filthier. The wards smelled like sewers and were terribly overcrowded; men suffering from highly infectious diseases such as typhus and dysentery had to share mattresses; and doctors had to walk over these mattresses to get to patients: 'When a man dies the next comer is put straight on to the same dirty mattress, between the same loathsome sheets'. Inevitably, infectious diseases such as typhus were rampant at the hospital, the report stating that twenty-five doctors at the hospital had died of fever since the war began. It adds that conditions at a hospital in Valjevo, a city in western Serbia, were even worse: 'More than 3,000 men lie there untended'.

Such descriptions suggest that little had changed in some hospitals since the time of the Crimean War when hospitals were equally insanitary, at least until English nursing pioneer and hospital reformer Florence Nightingale (1820–1910) and her team of nurses arrived in Turkey. Her observations of conditions at the Barrack Hospital in Scutari, a suburb of Instanbul, are recorded by British historian Cecil Woodham-Smith (1896–1977) in her biography of Nightingale.[14] The water supply for much of the hospital passed through the decaying carcass of a horse and was stored in tanks in a filthy courtyard next to open privies designed to cope with men suffering from diarrhoea. However, there was no means of flushing or cleaning the toilets and every breeze and puff of air blew poisonous fumes from the privies into the hospital's corridors and wards. Wooden shelves that ran round the wards also harboured rats for which the hospital was notorious.

Sick and wounded prisoners in some First World War prison camps experienced similar, if not worse, fates. For example, a report on prisoners in German camps, published in 1916, refers to them being rampant with typhus.[15] The report recounts the experiences of a Dr Ribadeau-Dumas, a physician in a Parisian hospital, who was taken prisoner and sent in February 1915 to a prison camp at Stendal in Germany. The German doctors stitched up wounds with a mattress needle and 'unasepticised' thread without washing their hands.

Many refugees and civilians caught up in the fighting fared little better. In her diary of 1914–1918, Englishwoman Florence Farmborough (1887–1978), who worked as a Red Cross nurse with the Imperial Russian Army during the First World War, describes an encounter with refugees at the village of Chertoviche on the Russian Front at the end of December 1915.[16] Some refugees stayed with the peasants in their huts while countless others huddled like sheep in old dugouts in the neighbouring woods: 'Fir-branches were the only protection from the mud; on them men, women and children squatted and slept, and the oozy slime gurgled and slopped beneath them'. On 2 January 1916, she reports that doctors in her unit went to visit an invalid soldier in a remote corner of the village and found no fewer than seventeen soldiers in one small hut. Some of the poorer peasants' huts were totally lacking in hygiene; pigs and hens also lived in the huts during the winter and 'spread dirt and discomfort on every side'. Flies and insects were a particular problem for Farmborough. On 6 July 1915, she describes wading through mud a foot deep infected by 'multitudinous flies and other highly obnoxious insects'. The insects found their way into tea, soup and beds: 'We killed them in their thousands and in their thousands they resuscitated'.

Microbes galore

The dirt and squalor endured by sick, wounded and also healthy troops, prisoners, refugees, civilians, as well as the doctors and nurses who provided medical care, proved ideal breeding grounds for infectious pathogenic microorganisms, or microbes as they are also called, and opportunities for the spread of a wide array of infectious diseases.

The vast majority of microbes we encounter in our daily lives are non-pathogenic and live in harmony with the human body. Each and every one of us co-habit with numerous bacteria, fungi and other microbes – in our guts, on our skin and elsewhere throughout our bodies. These microbes are parasitic as they depend on us for their existence. Most of them are harmless and some perform useful functions; for example, bacteria in our gut convert food into energy and essential organic chemicals such as vitamins.

'There are more bacteria hanging out just in the palms of your hands than there are humans on Earth', comments Sarah Everts, European Science Editor of *Chemical & Engineering News*.[17] The microbes we live with outnumber the cells in our body by a factor of ten and are known as human flora or the 'microbiome'. Some microbes help our immune system to fend off or destroy infectious agents including viruses and pathogenic bacteria before they infect the healthy cells in our bodies.

Viruses spread by hijacking living cells and multiplying inside them, before the newly formed virus particles then burst out of the cells and infect other living cells. Viruses, however, cannot replicate outside their host cells, while bacteria do not require living cells as hosts in order to exist and multiply. Most bacteria are single-cell organisms and are able to replicate in body fluids, multiplying in vast numbers so long as they are surrounded by an environment that can supply the nutrients they need. Bacteria are found virtually everywhere in nature: in humans, animals, plants, soil, water and air. Pathogenic bacteria cause disease in a variety of ways, with some able to multiply rapidly and produce poisons as soon as they enter the blood or gain access to body tissue. Viruses and pathogenic bacteria are not the only agents of infectious disease, however. Other agents include certain species of fungi and parasitic species of free-living, single-celled animals known as protozoa.

Pathogenic bacteria, viruses and other infectious agents spread from a source, such as soil or faeces, to humans, and from humans to humans in a variety of ways. They can be transported in air, dust, water, food and faeces, or carried by vectors, usually insects such as flies, fleas, ticks and lice, while contagious diseases are transmitted by bodily contact.

Common First World War infectious diseases

Dysentery

Bacilli 'have won more victories than powder and shot' commented eminent Canadian physician Sir William Osler in an address at the outbreak of the First World War.[7] Osler, who was Regius Professor of Medicine at Oxford University, was talking to British soldiers billeted at a camp near the village of Churn in Oxfordshire. The bacilli to which he referred are pathogenic, rod-shaped bacteria that cause diseases such as dysentery, or more specifically bacillary dysentery.

Bacillary dysentery is caused by bacilli known as *Shigella* and so the disease is sometimes called 'shigellosis'. Other infectious agents can also cause dysentery, the most important of which are parasitic protozoa known as *Entamoeba histolytica* which invade and destroy the tissues of the walls of the intestines. The resulting disease is known as amoebic dysentery or intestinal amoebiasis. Amoeba are single-celled animals with amorphous, jelly-like features.

The infectious agents of dysentery are found in faeces and are spread by food or water contaminated by the agents, the so-called faecal-oral route of infection. The agents are picked up and carried by unwashed hands that have come into contact with faeces in soiled ground or on dead bodies. The agents are also transmitted by insects, as T. J. Carey Evans, a captain in the Indian Medical Service, notes in an article published in 1917.[18] Following several years of experience of active service conditions he concluded that the common house fly was the main source of bacillary dysentery infection. Dysentery, whether bacillary or amoebic, is an acute inflammation of the large intestine that results in diarrhoea, typically with mucus and blood in the faeces: 'It is one of the most terrible of camp diseases, killing thousands, and, in its prolonged damage to health, one of the most fatal of foes to armies', Osler remarked in his address.

Treatment of dysentery varied during the war. Carey Evans advised 'absolute rest in the recumbent position' and a 'very sparing' diet that should initially consist of just barley water or rice water. All cases of acute bacillary dysentery, he suggested, should be treated every two hours with magnesium sulfate, an inorganic chemical compound containing magnesium, sulfur and oxygen. Its hydrated form is commonly known as Epsom salt. If there was no improvement, he recommended that patients be given bismuth salicylate, also known as bismuth subsalicylate. The compound, a salt of the metallic element bismuth and the organic compound salicylic acid, is an anti-diarrhoeal and anti-emetic drug. The treatment was then completed with another course of magnesium sulfate. Carey Evans reported that within fourteen days of this treatment in a field ambulance at Anzac Cove on the Gallipoli peninsular, 90 per cent of the Indian troops suffering from dysentery were returned to duty.

Soldiers suffering from amoebic dysentery were injected with emetine, an anti-protozoal drug that induces vomiting. The drug is one of numerous nitrogen-containing organic compounds, known as alkaloids, with complicated chemical structures that are found in plants. Emetine is extracted from the roots of the flowering plant *ipecacuanha* and was first obtained in its pure form in 1887, although impure crystals of the compound had been obtained as early as 1817.[19]

The use of emetine to treat amoebic dysentery during the war proved to be 'disappointing', although it had been effective in India before the war according to Carey Evans. Subsequently, a combination of emetine and bismuth iodide was used to treat the disease. Bismuth iodide, also known at the time as bismuthous iodide, contains bismuth and the halogen iodine in the ratio one to three. The combination of the two compounds was given orally and was 'far in advance of emetine by the needle', notes George C. Low, a temporary captain in the Indian Medical Service, in a letter published in the *British Medical Journal* in 1918.[20]

A range of other medicinal products were also employed to treat dysentery and the resulting diarrhoea during the war. They included another inorganic bismuth compound, bismuth carbonate,[21] which was used for convalescent cases after arrival in Britain. In Germany castor oil was also used to treat the condition.[22]

Enteric fevers: typhoid and paratyphoid

Troops without a supply of clean drinking water but eager to make a brew of hot tea scooped water from 'the top of waterlogged ditches and trenches without investigating what horrors of ordure or decay might lie in the murk beneath'. The description of the plight of soldiers on the Western Front shivering in the bitter winter of 1914/1915 comes from First World War historian Lyn

MacDonald in *The Roses of No Man's Land*: 'The result was an outbreak of enteric and dysentery'.[23]

Enteric is a synonym for enteric fever, an infectious intestinal illness. The most serious form of the fever is typhoid or typhoid fever as it is sometimes called. The fever is caused by pathogenic bacilli known as *Salmonella typhi*. The fever also has three other forms known as paratyphoid A, paratyphoid B and paratyphoid C. They are caused by *Salmonella paratyphoid* bacilli known as A, B and C respectively. The symptoms of typhoid include fever, headache, cough, sweating, chills, aching limbs, diarrhoea and a rash of red spots on the abdomen and back. Paratyphoid has similar symptoms but they are generally less severe and shorter in duration.

The disease is transmitted through food, milk or water contaminated by the urine or faeces of infected patients or carriers. 'Fingers and flies' are two of the chief factors in conveying the infection, Osler told members of the Society of Tropical Medicine and Hygiene on 20 November 1914.[24] The bacilli can survive in a symptomless carrier for months or even years after clinical recovery from the infection and then be transmitted every so often in the faeces or sometimes in the urine of the carrier. A report in December 1914 notes that it was quite impossible for soldiers in trenches previously occupied by other troops 'to keep their hands, clothes or general belongings free from mud and muddy water' that might be infected with typhoid and paratyphoid bacilli.[25]

Paratyphoid B accounted for just over 80 per cent of the cases belonging to the 'enteric group', noted Captain J.M. Fortescue-Brickdale at the fourth meeting of the Military Medical Society held in France on 27 October 1915.[26] The society, according to a report in the *British Medical Journal*, had 'evolved' the previous month when a 'few enthusiastic spirits' representing a group of hospitals in a district of France had gathered together 'under the trees by the banks of a river where Napoleon sat to plan his invasion of England'.[27] 'True typhoid' accounted for another 10 per cent, 4 per cent were paratyphoid A cases, while about 3 per cent were placed in the enteric group. although there was insufficient clinical or bacteriological evidence for the salmonella strain to be identified.

An account of typhoid fever in the German and Austrian armies reports sporadic outbreaks of the disease in October and November 1914, but fewer cases in December of that year.[28] It notes that the fever commonly disappeared spontaneously in the late autumn and winter.

Treatment of the fever, the report continues, required 'careful nursing, a fluid diet, and the free use of chemical antipyretics'. Antipyretics are drugs that reduce fever by lowering body temperature. Some analgesic drugs, such as aspirin, exhibit antipyretic activity. The use of aspirin, also known as acetylsalicylic acid, as a pharmaceutical drug was well established by the start of the First World War. It was discovered in 1853 by French chemist Charles Gerhardt (1816–1856) who synthesised an impure version by mixing the organic chemical acetyl chloride with sodium salicylate, the sodium salt of salicylic acid. In the late 1890s, the German firm Bayer developed a process for manufacturing the drug and by 1899 it was selling around the world. It was soon to become the world's best-selling drug.[29]

Cholera

Cholera is an acute infection of the stomach and intestine, in other words a gastrointestinal infection, and is caused by a bacterium known as *Vibrio cholerae*. The disease, which often occurs in epidemics, is contracted by consuming food or drinking water contaminated with infected faeces or vomit carried by flies or unwashed hands. It is characterised by the sudden onset of severe diarrhoea, followed by vomiting. The loss of fluid and sodium and potassium electrolytes from the body results in rapid and potentially fatal dehydration and cramp-inducing electrolyte imbalances. Patients are treated by replacing the fluids and electrolytes with salt solutions either through the mouth, known as oral rehydration therapy, or in severe cases by intravenous injection.

The treatment of cholera by intravenous injection of hypertonic saline solutions was pioneered before the First World War by Englishman Leonard Rogers (1868–1962), professor of pathology at the Medical College, Calcutta, India. Hypertonic saline solutions have higher concentrations of salt than normal saline solutions and when injected into a vein, hypertonic saline reverses the loss of fluid and salts from blood. Rogers showed that injection of a hypertonic solution of sodium chloride and small amounts of other chloride salts substantially improved the chances of recovery compared with injection of normal saline solutions.[30]

Many outbreaks of cholera occurred in the First World War. For example, in May 1915 Farmborough records a batch of six cholera patients arriving in her surgical unit on the Russian Front: they were in 'dreadful pain … Sometimes the cramp would draw them up into hideous contortions and they would writhe and twist in agony'.[31] The nurses massaged their legs, provided them with 'medicinal drinks' and then arranged for their transfer on ambulance vans to the nearest cholera hospital. 'Cholera and typhoid have taken toll of many, both military and civilian', she observes in a diary entry over a year later.[32]

The disease also occurred on many other fronts during the war. For instance, Lieutenant Colonel Sir Percy Lake, who commanded the India Expeditionary Force (IEF) in Mesotopotamia, reported an outbreak of cholera in April 1916 at the front on the River Tigris that flows from Turkey through the centre of Iraq.[33] Fortunately, the disease was rapidly brought under control.

Gas gangrene

On 23 August 1914, the BEF clashed with the German First Army at Mons, a town in Belgium near the border with France. The battle was one of the actions of the Battles of the Frontiers that the Allies fought to halt the German advance through Belgium and into France. Although the rapid and accurate fire of the professional BEF soldiers inflicted heavy losses on the Germans, the BEF was ordered to retreat for tactical reasons and because the force was heavily outnumbered.

In the early part of the war, there were no hospital trains and so it frequently took several days for the wounded to reach base hospitals.[34] Some wounded were taken to a hospital that had been hastily set up in a hotel at Versailles, near Paris. Macdonald notes in her book, *The Roses of No Man's Land*, that almost all of the wounds were infected with gas gangrene, an infection caused not by toxic war gases but by bacteria in the soil.[35] Macdonald quotes Geoffrey Keynes (1887–1982) of the RAMC, who was a surgeon at the hospital. He describes men dying like flies from appalling gas gangrene infections as wounds such as compound fractures became full of mud from heavily manured soil in France and Belgium, providing ideal sites for the bacteria to thrive and multiply. The wounds were masses of putrid muscle swelling up and rotting with the infection.

The RAMC had no previous experience of tackling the infection as gas gangrene had not been a problem fifteen years earlier when the British Army fought in South Africa, nor was it a significant problem on other fronts and theatres of the First World War. The gas gangrene bacteria 'present in highly dunged and cultivated soil' of the Western Front were absent from that of the South African veldt or the sun-dried sand of Egypt, Palestine and Mesopotamia, observed Sir Anthony Bowlby (1855–1929), consulting surgeon to British Army in France, in a lecture delivered before the Royal College of Surgeons of England on 14 February 1919.[34] Wet, cold weather and mud favour gas gangrene 'much more than heat and dust' and the infection was seen rarely in Gallipoli, Salonica and on the Italian front, while on the uncultivated land in Eastern Europe, the bacteria were present only in small numbers and 'apparently less virulent'.

Gas gangrene is the death and decay of wounded tissue infected by several species of clostridial bacteria, in particular the bacterium *Clostridium perfringens*, which are found in the soil and elsewhere. Clostridia are anaerobic bacterial organisms, meaning that they do not require oxygen to live and multiply. They do not infect healthy tissue but destroy damaged tissue and the surrounding tissue by releasing toxins. At the time of the First World War *Clostridum perfringens* was known as *Fränkel's bacillus* after the German bacteriologist Eugen Fränkel (1853–1925) who discovered the bacterium in 1892.

Gangrene itself is the death and withering of body tissue caused by restriction or cessation of the blood supply. If no infection is involved, it is known as 'dry gangrene', but if caused by bacterial infections then it is known as 'moist' or 'wet gangrene'. If gases are produced inside the infected body tissue as the tissue decays, the infection is known as 'gas gangrene'.

Little was known about the nature of the gases produced by the tissue decay until 1995 when scientists at the National Cheng Kung University Hospital in Taiwan reported a study of a fatal case of clostridial gas gangrene in a 60-year-old female diabetic patient.[36] The gangrene was caused by *Clostridium septicum*, a species of clostridial bacteria that resides in human gut flora but does not affect healthy tissue. The Taiwanese scientists' analysis of gas samples taken from the patient's damaged muscle showed that the gas was composed of 74.5 per cent nitrogen, 16.1 per cent oxygen, 5.9 per cent hydrogen and 3.4 per cent carbon dioxide.

In the First World War, surgeons treated gas gangrene cases by removing foreign material from the wound and excising the diseased tissue in a process known as 'debridement'. Unfortunately, in many cases even amputation of a wounded limb failed to halt the rapid advance of the infection.

Alternatively, and if the infection was less advanced, some surgeons attempted to inhibit the growth of the anaerobic bacteria by washing the wounds with antiseptics such as tincture of iodine, hydrogen peroxide solution or a solution containing hypochlorous acid. Tincture of iodine is a solution of iodine and an iodide salt, typically potassium iodide, in a mixture of ethanol and water. Hydrogen peroxide is a strongly oxidising colourless liquid made up of two atoms each of hydrogen and oxygen. Some surgeons also adopted procedures for treating gas gangrene that incorporated the intravenous injection of hypochlorous acid[37] or the injection of infected tissue with hydrogen peroxide.[38]

Bowlby noted a dramatic drop in the number of gas gangrene cases towards the end of the war. In spite of many thousands of operations being performed at the front during the Battle of the Somme in 1916, there were numerous bad cases of gas gangrene in the casualty clearing stations and base hospitals. Yet 'the greatest surgical epidemic of wound infection' that had ever been recorded was not repeated during the 1918 Battle of the Somme when Allied troops fought over the very same ground. Much of the land, Bowlby observed, was no longer freely manured and covered with rich crops as it had been in 1916. By March 1918, when the Germans launched their Spring Offensive, the once highly cultivated arable land was little more than 'uncultivated, unmanured, uncropped' desert, abandoned by man and animal alike, pitted with shells-holes, scarred by innumerable trenches and gunpits, and scattered with chalk subsoil on its surface. Gas gangrene bacteria was therefore unable to thrive.

Tetanus

On 23 July 1916, nurse Florence Farmborough writes in her diary that she had noticed 'the rapid development of both gangrene and tetanus resulting from flesh-wounds' of soldiers fighting on the Russian front.[39] She observes that enemy shrapnel shells were often filled not just with bullets but also with nails, particles of rusty metal and shrapnel fragments picked up from the battlefield. Many of these metal items were covered with filth. As a consequence, even small flesh wounds such as a scratch or graze could prove fatal.

Tetanus, also known as 'lockjaw', is an infectious disease caused by *Clostridium tetani*, an anaerobic species of bacteria found in cultivated soil, manure and even in dust contaminated with faecal matter. *Clostridium tetani* bacteria multiply in contaminated wounds at the site of infection, producing powerful toxins. After a few days, victims begin to suffer from muscle stiffness and then spasms and rigidity, initially in the jaw and neck and then in other parts of the body.

To treat the infection, the wound is first cleaned, then dead tissue, surrounding tissue and foreign matter are removed by debridement. During the First World War, wounded troops were injected with tetanus antitoxin, where available, both as a prophylactic measure to stop the disease developing and as a therapeutic measure if the symptoms of tetanus began to appear after surgery.

An antitoxin is an antibody produced by body cells to counteract toxins produced by invading bacteria. At the beginning of the war, tetanus antitoxin, or anti-tetanic serum as it was also called, was frequently produced from horses, according to bacteriologist Alfred MacConkey in a report on the topic published in 1914.[40] The animals were first immunised against tetanus by injecting them with tetanus toxin extracted from a broth of the tetanus bacteria and the nutrients they required. The immunisation stimulated the production of the antitoxin in the horses' blood serum, the fluid component of their blood. 'Anti-tetanic serum was often strikingly beneficial, provided large doses were given', notes Dr Pribram, one of the doctors who treated wounded soldiers in Germany.[41] Radical excision of all dead tissues and neighbouring freely bleeding tissues was the best treatment, as the use of antiseptics and cautery often failed to prevent the spread of infection.

Muscle spasms were relieved by administering chloral hydrate, a sedative and hypnotic drug prepared by the reaction of chlorine and an acidic solution of ethanol. The chemical was first synthesised by German chemist Justus von Liebig (1803–1873) in 1832. The spasms were also treated by injection of solutions of magnesium sulfate or a barbiturate, that is a drug derived from the nitrogen-containing organic compound barbituric acid.

A study published in 1918 compared the number of tetanus cases among wounded soldiers treated in British military hospitals during each year of the war. It showed that the number of cases was relatively high in 1914 but then dropped significantly in the years 1915–1918.[42] Furthermore, the death rate among tetanus cases dropped from almost 60 per cent at the start of the war to around 20 per cent from June 1917 onwards. These dramatic falls were attributed to the prophylactic injections of tetanus antitoxin introduced in October 1914.

Trench nephritis

In 1915, medical officers on the Western Front identified 'a new disease occurring amongst soldiers' who had spent some considerable time in the trenches.[43] They called the disease 'trench nephritis' or in some cases 'war nephritis'.

Nephritis is inflammation of the kidneys. The symptoms of trench nephritis observed in 1915 included a slight fever, headaches, tiredness, blood in the urine and oedema (swelling caused by accumulation of fluids in body tissues). The disease accounted for 5 per cent of medical admissions and more than 10 per cent of military hospital bed occupancy during the war, according to a report in 1986.[44] In another report, published in 2006, Robert Atenstaedt, an expert on the history of medicine, points out that the disease 'was a serious problem for the Allies, leading to 35,000 casualties in the British and 2,000 in the American forces.'[45] He adds that the disease led to hundreds of deaths.

The cause of the disease was the subject of much discussion during the war. Some doctors thought it was caused by a bacterial infection, while others attributed the disease to exposure to the cold conditions in the trenches. Tin poisoning from tinned foods or chlorine poisoning from the hypochlorites used to purify water were also mooted as possible causes.[46] According to Atenstaedt, no definite cause for the disease was ever discovered, although it is now thought that it may have been induced by a hantavirus – a type of virus carried by rodents.

Patients were treated with bed rest, a fixed diet of milk, bread, rice, butter, potatoes, greens, jam and occasionally fruit, observes J. Michell Clarke in a paper on trench nephritis published in August 1917.[47] Clarke, an RAMC medical officer and professor of medicine at the University of Bristol, notes that pure crystallised sodium carbonate in water was administered hourly as a diuretic to patients who exhibited deficient urinary excretion when they were admitted. If their blood pressure was high, they were given nitroglycerine, the same chemical that was used to make cordite. Adrenalin, a nitrogen-containing organic chemical related to benzene, was given to patients with low pressure.

Typhus and other louse-borne diseases

In early 1915 an outbreak of typhus spread like wildfire through Serbia. Whole villages, including one of 2,000 inhabitants near Skopje (now the capital of Macedonia), were 'wiped out' by the disease, according a report in the *British Medical Journal* in May 1915.[48] The epidemic was thought to have originated among a group of Austrian prisoners at Valjevo.[13]

The disease killed not just civilians, prisoners and troops, but also the doctors and nurses who tended them. They included Elizabeth Ross, who had trained as a doctor in Scotland, spent many years working in Persia, and worked as a ship's surgeon along the coasts of China and Japan.[49] When war broke out, she went to Serbia at the request of the Russian Army Medical Service. She took charge of the typhus wards at the military hospital in Kragujevac but contracted the disease there and died. According to the May 1915 *British Medical Journal* report, 360 Serbian doctors died, and all but one of the six American surgeons and twelve nurses who went in Red Cross units to Serbia succumbed to the disease.

Typhus epidemics also occurred elsewhere in Europe during and immediately after the war; for example, in Germany in the spring and summer of 1915,[50] and in Poland where 40,000 people died.[51] The disease was endemic on the Eastern Front but the armies on the Western Front were spared its ravages, remarks Oxford University professor of the history of medicine Mark Harrison in an account of 'war and medicine in the modern era'.[52] Typhus killed 150,000 soldiers in Serbia during the first six months of the war and took 'three million Russian lives by the end of the war', according to British science writer Michael White.[53]

Typhus is any one of a group of acute infectious fevers that typically occur in overcrowded and insanitary conditions. The disease, which should not be confused with typhoid, is caused by spherical or rod-like parasitic bacterial microbes with some characteristics of viruses. These organisms, known as 'rickettsia', are found in and transmitted to humans and animals by lice, ticks, mites and fleas. The symptoms of the disease include a prolonged high fever, severe headache, widespread rash and delirium.

The typhus infection responsible for the severe epidemics in the First World War is known as 'epidemic typhus' or 'louse-borne typhus'. It is caused by *Rickettsia prowazeki*, a species of rickettsia transmitted by the infected faeces of the common human body louse, *Pediculus humanus*. As soon as the rickettsial organisms enter the victim's blood, they start to multiply.

Typhus was not the only louse-borne disease to occur in Europe during the war. In March 1915, Scottish physician William Hunter (1861–1937) arrived in Serbia with the British Military Sanitary Mission to discover that a combined epidemic of typhus and relapsing fever was raging in the country. Relapsing fever is an infectious disease caused by a species of pathogenic bacteria known as *Borrelia recurrentis*. The bacteria are also transmitted by human body lice. The symptoms of the disease, which include fever, chills, vomiting and muscle pain, appear suddenly and then continue for several days before disappearing. They then recur about a week later. The mortality rate can be as high as 40 per cent.

In a report published after the war, Hunter observes that the relapsing fever epidemic had broken out in December 1914.[54] The number of patients suffering from the disease in Serbian military hospitals had risen from 2,200 in January 1915 to 8,500 in April that year. The incidence of the disease in the civilian population was not known but likely to be at least five times greater. He suggests that the number of typhus and relapsing fever cases in the first three months of 1915 in Serbia, with its population of three million, fell little short of 500,000 cases with at least 120,000 deaths.

Hunter points out that the typhus and relapsing fever epidemic was widespread throughout Serbia, with every town, village and hamlet suffering. Furthermore, the military hospitals in the country provided ideal breeding grounds for the spread of the lice as the conditions in the hospitals were indescribably bad:

overcrowded, and sanitary and disinfection arrangements were non-existent. Similarly, the carriages of trains in the country were overcrowded with 'peasants in their sheep skin coasts – the friendly homes of countless lice'. The country's railway system acted as 'rivers of infection conveying infection up and down the country – from the army areas to the civilian areas, and from the latter back to the army'.

Strategies for treating typhus and relapsing fever patients during the First World War involved their isolation, and the delousing of their bodies and clothes to prevent the infection spreading. Various drugs, if available, were used to treat the symptoms of the disease. For example, a report on the treatment of typhus in Serbia during the epidemic in the early months of 1915 refers to digitalis being 'given by the mouth and hypodermically' to patients, with strychnine also being administered hypodermically to combat the disease.[55]

Digitalis is extracted from the dried leaves of plants in the foxglove family. It contains various organic chemicals known as glycosides that stimulate the muscles of the heart and is therefore used to treat heart disorders. Strychnine, like the emetine used to treat amoebic dysentery, is a type of organic chemical known as an alkaloid. The highly toxic chemical occurs in abundance in the seeds and bark of many plants. During the war it was used in small doses as a tonic to stimulate nerves and muscles.

Quinine, another alkaloid, was also used, according to K. David Patterson, a professor at the University of North Carolina at Charlotte, USA.[56] The alkaloid is obtained from the bark of the cinchona trees that grow in the tropical forests of South America. It was used in small doses as a tonic and to reduce fever, and in larger doses for the treatment of malaria.

Although typhus was a scourge all along the Eastern Front during the First World War, it was contained by both sides along the Western Front. The stringent precautions taken by army sanitary units on the Western Front to prevent the spread of lice could possibly account for the total absence of typhus on the front. Even so, these measures did not prevent the occurrence of another louse-borne disease on the front, often known as 'trench fever'.

Trench fever was transmitted by human body lice infected with a bacterial microorganism once known as *Rickettsia quintana*, but now called *Bartonella quintana*. Symptoms of the disease included high fever, severe headaches, inflamed eyes, leg pains, pink rashes and shivering. The fever typically lasted for about five days followed by remission and then recurrence after another few days. In some cases its recurrence-remission cycle occurred several times.

The disease was discovered among soldiers in the trenches of the Western Front in the spring of 1915 notes Sir Wilmot Herringham (1855–1936), a consulting physician to the British forces in France, in a report on the topic published in 1917.[57] Although the infection was relatively benign compared with typhus, it incapacitated hundreds of thousands of Allied troops for periods of three months or more during the war. Between 1915 and 1918, an estimated 450,000 British troops had suffered from the disease.[58] Herringham observed that patients might remain ill for many weeks, although no cases proved fatal. He reported that many drugs, including quinine, were used to treat the infection but 'none had any effect upon the course of the fever', except aspirin or morphine which relieved the pain temporarily.

Trench foot

In the cold and wet winter of 1914/1915, many troops on the Western Front had to walk, stand or sit in waterlogged and poorly drained trenches with their non-waterproof boots immersed in a soup of mud and water. Inevitably, the footwear became sodden with mud and their feet cold and damp. In some cases their feet remained wet for days on end. This prolonged exposure led to a medical condition known as 'trench foot'. It was characterised by numbness, swelling, lesions, blackening of the toes, the skin on the feet dying and peeling off, and other symptoms.

To complicate matters, pathogenic microbes that thrived in the mire of the trenches often seized the opportunity to infect the damaged feet.

Early stages of the disorder were typically treated by washing the feet with a medicated soap containing antiseptic chemicals such as boric acid and the organic compound camphor. The skin was then painted with a solution of an antiseptic such as picric acid. Aspirin, or some other analgesic, was used to relieve pain. In all cases, patients were injected with anti-tetanus serum. In advanced cases, which occurred if the condition was left untreated, gangrene set in and surgeons sometimes found it necessary to amputate one or more toes to prevent the infection spreading.

By 1916, it was well recognised that preventative as well as curative measures were needed to tackle trench foot. Preventative treatment included the provision of knee-length rubber boots, improvements in diet, and rubbing the feet with grease and whale oil.[59] The Inter-Allied Surgical Conference that took place at Val-de-Grâce, Paris, on 11–16 March 1916, concluded that preventative treatment should also include improved hygiene, draining trenches, dry and warm shelters, and frequent reliefs.[60] In addition, troops should dry, clean and massage their feet, and change their socks each day.

Malaria

Troops fighting in the warm climates of East Africa, Macedonia, Palestine, Mesopotamia and other tropical and subtropical theatres of the First World War, faced not just the dangers of infectious illnesses such as dysentery but also those of tropical diseases, most notably malaria. From early January to the beginning of May 1917, for example, over 20 per cent of the British-led force in East Africa were admitted to hospital with malaria.[61] Furthermore, between January and mid-May the following year, around 90 per cent of the men in some units fighting on the Macedonian Front contracted the disease.

Malaria is an infectious disease caused by the presence of various species of parasitic protozoa, known as *plasmodia*, in red blood cells. The parasitic organisms are transmitted from human to human by female *Anopheles* mosquitoes. These small, winged, blood-sucking insects acquire the parasites when they feed on the blood of people infected with the disease. The organisms multiply inside the mosquitoes and are then injected into the blood of another victim when the insect bites the individual. The parasites then migrate to and multiply in the liver and other organs. The infection results in shivering, fever, sweating, convulsions, vomiting, anaemia caused by the loss of healthy red cells, and a number of other symptoms.

Various protective items, procedures and chemicals were employed during the war to prevent the spread of the disease. They included the use of mosquito nets, and the draining of swamps, ditches and other standing waters where the mosquitoes bred. A British memorandum on diseases in the Mediterranean war area, published in 1916, refers to the use of mosquito repellents such as oil of bergamot and oil of cassia.[62] These two oils are found in aromatic plants and known to deter insects. Oil of bergamot is a fragrant, yellow-green volatile mixture of organic chemicals extracted from the rinds of sour citrus fruits known as bergamot oranges. The most valuable component of the oil is linanyl acetate, a compound that is also found in lavender oil and is used to make perfumes. Cassia oil is a yellow or brown liquid with a cinnamon-like smell obtained from the bark of the evergreen tree *Cinnamomum cassia*. The oil is a mixture of organic chemicals, notably cinnamaldehyde.

The most important chemical used in the First World War to tackle malaria was another organic compound extracted from the bark of a tree. That chemical was quinine which, as we saw above, was also used to treat typhus. Quinine is a 'base', meaning that it combines with acids to form salts. During the war, quinine was administered prophylactically and therapeutically as a salt, for instance as quinine sulfate or quinine dihydrochloride.

Quinine salt tablets were given daily to troops fighting in countries such as Macedonia as a prophylactic measure to prevent the development of the infection.[63] Solutions of a quinine salt were also administered as drinks to troops diagnosed with malaria. The treatment was usually preceded by a dose of calomel, an inorganic mercury- and chlorine-containing chemical compound otherwise known as mercurous chloride or mercury(I) chloride. This naturally-occurring compound was widely used as a laxative. If the quinine drug failed to halt the infection when given by mouth, the doctors resorted to intramuscular or intravenous injections of a quinine salt solution.

Sexually transmitted infections

In *Old Soldiers Never Die*, Richards describes a queue of about 300 men, the majority of them 'lads', waiting their turns to go into the 'Red Lamp', a brothel in Bethune, France.[64] Robert Graves refers to the same brothel in *Goodbye to All That*.[65] He notes that the Red Lamp brothels were used by men whereas the Blue Lamp brothels at Amiens, Abbeville, Le Havre, Rouen and all the large towns behind the lines, were reserved for officers.[66] Later in the book Graves points out that the boys, as he calls them, had money in their pockets and knew that they might well be killed within a few weeks: 'They did not want to die virgins'.[67]

'Syphilis and gonorrhea are the industrial diseases of prostitution' observes a correspondent in a letter published in a 1917 issue of the *British Medical Journal*.[68] Venereal diseases occurred almost on an industrial scale in the armies fighting in the First World War and were responsible for a significant portion of the wastage of the war. The average stay in hospital for British and Dominion troops suffering from syphilis, gonorrhoea and other sexually transmitted diseases was around thirty-eight, twenty-nine and thirty-one days respectively.[69] A total of over 416,498 British and Dominion soldiers and officers were admitted to hospital for the treatment of venereal disease during the war according to one source.[70]

In a report on the management of venereal diseases in Egypt during the war, Australian doctor Sir James Barrett (1862–1945) notes that 2.5–3 per cent of the Australian soldiers waiting in the country in early 1915 to go the Dardanelles had contracted a venereal disease.[71] A total of 1,344 of these troops were sent back to Australia and another 450 to Malta for treatment. During the five months from January to May 1916, there were no fewer than 10,000 known cases of venereal disease among the soldiers in Egypt.

The picture was much the same in other armies. For example, about 4 per cent of the troops in the Austrian Army suffered from venereal disease, according to medical officers serving in the army.[72] Five per cent were infected at the front, 20 per cent along the lines of communication and the remaining 75 per cent contracted the disease 'outside the sphere of the army', with the majority of infections picked up in brothels.

Syphilis is a chronic sexually transmitted contagious disease caused by the bacterium *Treponema palladium*. The disease develops in several stages resulting in the formation of lesions throughout the body as well as other symptoms. Gonorrhaea, another sexually transmitted contagious disease, is due to infection by the bacterium *gonococcus*, also known as *Neisseria gonorrhoeae*. The symptoms include pain on urination and the discharge of pus.

During the First World War, arsenic-containing organic chemicals were widely used as drugs to treat syphilis. The most common was a pale yellow powdery synthetic chemical known variously as 'salvarsan', 'arsphenamine', or simply '606'. Salvarsan was typically dissolved in saline solution and administered to patients by intravenous injection. The arsenical was discovered in 1909 by the German scientist Paul Ehrlich (1854–1915) and his team, who had prepared hundreds of organic compounds containing arsenic and tested them for their anti-microbial activity. Compound number 606 proved highly effective in curing rabbits infected with syphilis. The drug was patented and first marketed under the tradename 'Salvarsan' in 1910. Before then, inorganic mercury salts had been commonly used to treat syphilis.

Ehrlich is regarded as the founder of the science of chemotherapy – that is the treatment of diseases with chemicals. Until the end of the nineteenth century, the vast majority of drugs were naturally-occurring chemical substances extracted from plants, quinine for example. Ehrlich showed that drugs could also be chemically synthesised to treat diseases. Ehrlich and Russian scientist Ilya Ilyich Mechnikov (1845–1916) won the 1908 Nobel Prize in Physiology or Medicine 'in recognition of their work on immunity'.

In 1912, Ehrlich discovered neosalvarsan, also known as neoarsphenamine or novarsenobillon. The arsenical has a deeper yellow colour than its chemical cousin salvarsan, but is less toxic and more soluble in water. Furthermore, it was easier to manufacture and administer to patients. Neosalvarsan therefore became the drug of choice for treating syphilis in the latter years of the war.[73] Both arsenical drugs, however, carried a high risk of chronic arsenic poisoning. In *Testament of Youth*, English writer Vera Brittain describes a syphilitic orderly dying in convulsions after an injection of salvarsan at a hospital in Malta in 1917.[74] The following year, three New Zealand medical officers reported two fatal cases of jaundice 'following injections of novarsenobillon' at the New Zealand General Hospital in Codford, a village in Wiltshire.[75]

Gonorrhoea was treated by irrigating the urinary tract with a solution of a urinary antiseptic such as urotropin, otherwise known as hexamine.[76] The same chemical was used in gas masks during the First World War to provide protection against phosgene. A dilute solution of the strongly oxidising inorganic compound potassium permanganate was also used as an antiseptic to wash the urinary tract. Another treatment involved packing the front portion of the urethra, that is the tube through which urine is discharged from the bladder, with a cylinder of gauze soaked in a solution of argyrol, the tradename for a mildly antiseptic silver-protein preparation.[77] The antiseptic was developed by American chemist Albert Barnes (1872–1951) and first marketed in 1902.

It was not until the late 1930s that sulfonamides, a family of synthetic nitrogen- and sulfur-containing organic compounds, came into use as antibacterial drugs for treating sexually transmitted infections. Nowadays, syphilis and gonorrhoea are typically treated with an antibiotic like penicillin. Antibiotics are chemical substances, derived from microorganisms such as fungi, which inhibit or prevent the growth of other microorganisms. Antibacterial sulfonamides are not classified as antibiotics as they are not produced by living organisms.

The 1918 Spanish flu pandemic

On 21 September 1918, the White Star Line's *Olympic*, a sister ship to the *Titanic*, arrived in Southampton with 146 officers and 5,828 other ranks of the AEF. They included 573 cases of influenza and pneumonia according to a report in the *British Medical Journal*. Within a week, a further 1,000 cases occurred.[78] Around 300 of the troops died. The *Olympic* was the first of the ships carrying troops of the AEF to Europe to have an influenza outbreak in the 1918 pandemic.

In his account of the war published in the book *Last Post*, Kenneth Cummins, a Royal Naval Reserve, remembers going down to the pier head in Liverpool and seeing corpses of men who had died from flu being carried off United States troopships.[79] In another account, Bill Stone, a seaman in the Royal Navy, notes that thousands of sailors died of flu in Plymouth: 'They'd be on parade and they'd just keel over'.[80]

According to some experts, the disease probably originated in the United States before it rapidly spread around the world. It was called 'Spanish influenza' because the uncensored press in Spain, which remained neutral during the war, publicised accounts of the pandemic – unlike the censored press of the belligerent nations. When the news from Spain reached other countries it was assumed that the disease had originated in the country.

Wartime Europe, with its movements of armies and troops living in close proximity to one another, proved fertile ground for the rapid spread of the highly infectious disease. According to some estimates, by the end of the pandemic between 50–100 million people worldwide had died from the disease.[81] That is roughly equivalent to wiping out the whole population of Great Britain – which currently stands at just over 60 million.

The pandemic was caused by the H1N1 virus: the 'H' and the 'N' refer to the two types of spike-like proteins protruding from the surface of the virus particle. There are sixteen different subtypes of the H protein, labelled H1 to H16. Similarly, the N protein has nine subtypes, labelled N1 to N9. Each strain of influenza has a different combination of H and N subtypes: the strains H1N1 (Spanish flu) and H5N1 (avian flu) are examples.

The first outbreaks of the disease in the British Army in France occurred in April and May 1918.[82] Symptoms of the disease included fatigue, fever, headaches and body pains. More ominously, however, was that victims often became cyanosed, a blue discoloration of the skin due to insufficient oxygenation of the blood, and bled from the mouth, nose and in some cases the ears. The onset of these symptoms was sudden and could result in death within days if not hours. Many victims subsequently became infected with bacterial pneumonia which increased the death rate even further.

Treatment varied immensely. Vaccination for both influenza and pneumonia cases was attempted with limited success using a vaccine consisting of pneumonia bacteria 'prepared from organisms isolated from the blood or lungs of the most toxic cases', according to a report in 1919.[83] Medical care focused principally on treating the symptoms rather than the cause of the disease. In a report published just after the Armistice, a medical officer working in the hospital at the Canadian Army camp at Bramshott, a village in Hampshire, describes putting each influenza patient on a fluid-only diet consisting of 'four pints and four eggs'.[84] Calomel and Epsom salts were administered as purgatives to remove poisons from the body; chest pain was relieved with mustard leaves, poultices and other preparations; coughs with balsamic inhalations; and chills, headaches, and other aches and pains with aspirin and Dover's powder – a sedative and sweat-inducing mixture of opium and the powdered dry roots of the ipecacuanha plant. Ammonium bromide, codeine, heroin and morphine were also given as sedatives. Strychnine, camphor in oil, adrenalin, atropine, various digitalis preparations, as well as brandy or whisky, were administered as stimulants. Patients were also dosed with urotropin, quinine sulfate, creosote, potassium iodide, atropine – an alkaloid extracted from the plant commonly known as deadly nightshade or belladonna – and a variety of other chemicals.

'From the multitude of agents employed their comparative futility is obvious', concludes the Canadian medical officer. He advocates taking *Primum non nocere* as a line of treatment. The Latin phrase, meaning 'first, do no harm', underpins the standards of medical ethics of the Hippocratic oath. Of the 2,247 cases admitted to the Bramshott hospital in September and October 1918, 163 (just over 7 per cent) died, with death occurring almost invariably from respiratory complications.

Preventing the spread of infectious diseases

In the First World War, vast numbers of troops lived and fought closely together in the squalor of the trenches and battlefields, and were exposed to a wide array of infectious diseases. So how was it that the wastage of war from disease, in terms of the percentage of lives lost, was much lower than in earlier conflicts such as the Crimean War and the South African wars? How was it that, for the first time in a major war, deaths from battle injuries exceeded those from disease?

A number of factors contributed, including vastly improved organisation and administration of medical and nursing care of the sick and wounded. During the Crimean War, orderlies were the only nurses in the British Army, while nurses such as Florence Nightingale were volunteers. And even at the time of the Boer wars in South Africa that took place between 1880 and 1902, the British government was reluctant to send trained nurses to the front. The Voluntary Aid Detachment in Britain, which provided nursing services in the field under the protection of the Red Cross during the First World War, was founded in 1909, and the Royal Army Medical Corps in Britain was established in

1898, albeit with just 800 regular medical officers. In August 1914, the RAMC had 4,807 officers and men, which, by the end of the war, had increased to 144,514.[85]

In Germany, there were '3,000 medical men attached to the Red Cross, 400 dentists, 1,800 chemists, and 92,000 men in the medical services', reported a German professor at the German Medical Congress held in Warsaw, Poland, on 1–3 May 1916.[86] There were a further 72,000 men in the medical services at home, and 22,000 volunteer nurses and 6,800 sisters on the lines of communication. The sick and wounded were taken from the field hospitals back to Germany in 238 hospital trains and in other trains 'for slight cases'.

The First World War saw not only improvements in the care and treatment of troops suffering from infectious diseases, especially the treatment of the symptoms of these diseases, but also significant developments in the prevention of these diseases. 'No infectious disease has ever been noticeably reduced, much less exterminated, by treatment; whereas a large number have been greatly reduced, if not entirely eliminated, by efficient preventive measures', commented three correspondents in a letter to the *British Medical Journal* in 1919.[87]

Preventative measures during the First World War included the provision of health education and the training of troops, especially with respect to hygiene, sanitation and the risk of catching a sexually transmitted infection. Improved equipment and methods of diagnosing and investigating infectious diseases enabled pathologists and bacteriologists to determine not only the nature and sources of infectious diseases but also ways of curbing them. From October 1914 onwards, for example, the British Army began to send motor vans fitted out as mobile bacteriological laboratories to work alongside casualty clearing stations or groups of these stations.[88] The laboratories were equipped with the instruments required to carry out bacteriological investigations, such as incubators for growing bacterial cultures and microscopes for examining the bacteria.

The incidence of infectious diseases like typhoid and tetanus was greatly reduced, especially on the Western Front, by the prophylactic inoculation of troops with vaccines and serums. Following the invasion of France, the Germans carried out an extensive programme of vaccination against typhoid in November and December 1914 and January 1915.[28] The programme cut the mortality from the disease to almost a half. By the end of the war, it became compulsory for all troops in the armies of the Central Powers to be inoculated against not only typhoid but also cholera.[89]

In Britain, the use of anti-typhoid vaccines was pioneered by Sir Almroth Wright (1861–1947). He began his investigations into typhoid fever in 1893 soon after he had been appointed as professor of pathology at the Army Medical College based at the Royal Victoria Hospital at Netley, near Southampton. In 1897, he showed that immunity against typhoid could be induced in human volunteers by injecting them with dead *Salmonella typhi* bacteria.[90]

In 1902, Wright was appointed professor of bacteriology at St Mary's Hospital, London, where he continued his work on typhoid vaccines. When Britain entered the First World War, the British Secretary of State for War, Lord Kitchener (1850–1916), ordered that all soldiers sent to the front should be vaccinated against typhoid. The Inoculation Department at St Mary's Hospital subsequently produced ten million doses of the vaccine.

According to British author Guy Hartcup, by the end of 1915 nearly 98 per cent of British soldiers had agreed to be inoculated against typhoid.[91] This large-scale programme was driven by Keogh who, as we saw above, was Director General of the Army Medical Service during the war. The typhoid vaccination programme cut the risk of contracting the disease to about one tenth. In January 1916, the British introduced a new triple vaccine against typhoid and paratyphoid A and B, known as the TAB vaccine.

Sterilisation of medical and surgical instruments, equipment and dressings also played a part in preventing the spread of pathogenic microorganisms. For example, the 1916 British memorandum on diseases in the Mediterranean war area advises that syringes and needles used for injections must be 'most carefully sterilised by boiling and by rinsing thereafter with 1 in 20 carbolic acid'.[62] Aqueous solutions of carbolic acid, a coal tar chemical also known as phenol, were widely used in the war as disinfectants and antiseptics.

Improvements in sanitary arrangements also made a significant contribution to halting the spread of infectious diseases. In an article on bacteriology and sanitation science in the British Army, Atenstaedt notes that in the first few months of the First World War, each military unit had personnel responsible for carrying out sanitary duties.[92] Every base had use of a sanitary unit and every division was given a sanitary section. By the end of 1917, the Sanitary Service had 25,000 officers and men in France. Their responsibilities included: protection and purification of water supplies; building latrines; the provision of bathing facilities for troops; cleansing and disinfection of clothing, blankets and billets; and sewage disposal.

The role of chemistry in these arrangements was widely recognised. 'Chemistry, pathology, entomology and bacteriology are all four working' in the service of preventative medicine to keep troops healthy, comments a *British Medical Journal* correspondent in Northern France in 1916: 'The result is real and effective sanitation.'[93]

Clean water

According to one report on military hygiene published in 1915, a soldier 'in bivouac' required 1 gallon of water for drinking and cooking each day.[94] In barracks, on the other hand, the daily water requirement per man was 20 gallons, while the daily requirements for horses and cattle were 8 gallons and 6 gallons for mules. Chemistry played a crucial role in the provision of clean water to maintain an effective fighting force.

Wherever possible, the water was extracted directly from natural sources such as springs, streams and rivers; however, in many cases it was necessary to sink wells and borings, and to install pumping plants and mains. During the Battle of the Somme in 1916, for example, the British laid more than 120 miles of pumping mains to ensure an adequate supply of water to the troops. Bacterial examinations of samples taken from these sources determined the microbiological quality of the water. The samples were also tested for turbidity (cloudiness), colour, odour, solid residues, alkalinity, total hardness, and chemical components such as chloride and nitrate.

The army employed various methods to purify water: boiling was often used, although that required substantial amounts of fuel that were not always available; filtration through sterilised sand or porous clay removed solid impurities; and the addition of an alum, a chemical compound containing aluminium and sulfate, coagulated and precipitated impurities suspended in the water and therefore clarified it.

Various chemicals were used to sterilise water, including potassium permanganate and sodium hydrogensulfate, also known as sodium bisulfate. However, the favoured method was chlorination. Bleaching powder, also known as chloride of lime or calcium hypochlorite, was typically used for this purpose, although liquid chlorine was also used. However, not everybody felt happy with this method of sterilisation. While working in a casualty clearing station in the Somme region of France, Sister Mary Hall complains of being fed up with chlorinated water, as everything the nurses ate that was cooked in water 'seemed to taste of chlorine'.[95]

Disinfectants

Chloride of lime, that is calcium hypochlorite, was used not just to kill the microbes in water but also as a disinfectant in the trenches and elsewhere. 'A rat scampered across the tin cans and burst sandbags, and trench atmosphere reasserted itself in a smell of chloride of lime', writes British war poet and soldier Siegfried Sassoon in his *Memoirs of an Infantry Officer*.[96]

Other disinfectants were also used. For example, a 1916 report on personal hygiene refers to the use of a dilute solution of cresol in laundering troops' dirty underclothing.[97] Cresol is one of the many organic compounds extracted from coal tar; its chemical structure is closely related to carbolic acid. The same report notes that 'if vermin were detected', the clothes were also soaked for twenty-four hours in a dilute solution of mercury perchloride, known nowadays as mercuric chloride or mercury(II) chloride, which contains mercury and chlorine in the ratio 1:2. Disinfectant and insect repellent oils, such as lavender oil, were also sometimes used to disinfect the floors and walls of hospital wards.

A lousy problem

Of all the vermin encountered in the war, lice were among the most notorious. They infested trenches, dug-outs, billets, blankets, beds, and the troops' clothing and kit. They not only proved an immense irritation to troops but, as we saw above, also carried infectious diseases.

'We were lousy', comments British soldier Harry Patch (1898–2009) in the *Last Post*.[98] In the four months that he was in France, Patch never had a bath or clean clothes to put on, so it was no wonder that the clothes were infested with lice. Each louse had its own bite, its own itch and was the size of a grain of rice. Many troops attempted to burn the lice out by running the flame or hot wax of a candle along the seams of their clothes.

Just after the start of the war, Cambridge University zoologist Sir Arthur Shipley (1861–1927) suggested the application of a liquid such as petrol, paraffin oil or turpentine to a person's head to remove nits, that is head lice.[99] He also advised that clothing infested with nits and lice should be steamed, ironed with hot candle wax, or treated with petrol or paraffin.

In Russia during the war, hospital patients were deloused by shaving hair from the body and treating them with kerosene or benzene.[56] Steaming was widely used for disinfecting clothes, linen, possessions and buildings. Fumigants such as benzene fumes or sulfur dioxide, a gaseous compound, together with solutions of calcium hypochlorite, carbolic acid or other disinfectants, were also employed for lice disinfestation in buildings and railway carriages.

In addition, a range of insecticidal powders was used to tackle the lice problem. The report on personal hygiene in the war, for example, notes that a dusting powder of 'ammoniated mercury', zinc oxide and magnesium silicate was highly effective.[97] Ammoniated mercury, an inorganic chemical compound more accurately called ammoniated mercuric chloride, is a white odourless and poisonous powder consisting of one part each of the elements mercury, nitrogen and chlorine, and two parts hydrogen.

Another powder, widely used by the British Army, was known as 'NCI powder'. It was prepared from powdered naphthalene and small amounts of creosote and iodoform.[100] Naphthalene, an organic chemical extracted from coal tar, is better known as the main constituent of mothballs. Creosote is produced from wood tar or coal tar and is a mixture of carbolic acid, cresol and other organic chemicals. Iodoform, a disinfectant organic compound, is similar in chemical composition to chloroform. Each iodoform molecule contains one atom each of carbon and hydrogen, and three atoms of iodine.

Antiseptics

One of the saddest experiences of the First World War was to learn from those who had been at the front that 'practically all wounds were septic', remarked English surgeon Sir Rickman Godlee (1849–1925) at a discussion on 'surgical experiences of the present war', held in November 1914 at the Medical Society of London.[101] He enquired at the meeting what attempts were being made to disinfect wounds.

Godlee, a pioneer of brain surgery, was a keen advocate of antiseptics. He had not only studied and written about the topic but was also the nephew of Joseph Lister (1827–1912), the English surgeon who is most famous for introducing the use of antiseptics during surgery. Lister described the work in a paper 'on the antiseptic principles of the practice of surgery' that he read before the British Medical Association in Dublin in August 1867.[102] He also promoted the use of aseptic techniques, that is nursing and surgical procedures that prevent microorganisms coming into contact with and infecting wounds. Later in that same year, Lister published a series of three papers 'on the use of carbolic acid'.[103] He used the coal tar chemical not only to clean wounds, but also to scrub surgeons' hands, and to sterilise surgical instruments and post-operative dressings. The preparations typically consisted of a solution of carbolic acid in boiled linseed oil or of a putty-like paste made by mixing the solution with 'common whiting'. Whiting is a finely ground form of chalk or limestone, that is calcium carbonate or 'carbonate of lime' as it was known at the time.

Carbolic acid was subsequently widely employed as an antiseptic and disinfectant right up to and throughout the First World War. In particular, antiseptics became the first line of defence against infectious organisms in the treatment of wounds. Carbolic acid, the cresols, and related coal tar chemicals were also exploited as germicides in household commercial products introduced in the latter half of the nineteenth century and used during the war. These products include antiseptic coal tar soaps, such as Wright's Coal Tar Soap, and disinfectants such as Lysol and Jeyes Fluid.

Chemicals that kill pathogenic microbes are sometimes called 'germicides'. The term 'antiseptic' is generally used for chemical agents that are applied to skin and living tissue to prevent or treat sepsis, that is the invasion and growth of infectious microbes, especially bacteria, in the tissue. Disinfectants are germicides that destroy infectious microorganisms in or on non-living substances and objects such as dirty water, contaminated surgical instruments and soiled clothing.

A wide range of chemical substances were used as antiseptics to clean, wash, dress wounds and to treat the surrounding skin and tissue. They included tincture of iodine and aqueous solutions of inorganic compounds such as hydrogen peroxide, magnesium sulfate, mercuric chloride, mercuric iodide, silver nitrate or sodium hypochlorite. The boric lint used for dressing war wounds consisted of lint impregnated with boric acid, an antiseptic inorganic compound containing the chemical element boron. It was also known as boracic acid.

Organic substances were also used in antiseptic preparations: they included not only carbolic acid, but also picric acid and salicylic acid, both of which are chemically related to carbolic acid; ichthyol – an oil shale chemical also known as ichtammol or ammonium bituminosulfonate; quinine hydrochloride; glycerine; and paraffin. Picric acid is the self-same organic chemical that was used to manufacture the explosive lyddite.

Some antiseptic preparations used in the war were mixtures of chemicals. For example, in 1917, a surgeon at a war hospital in Northumberland reported the use of a novel type of antiseptic paste to treat large infected wounds. The surgeon, whose surname was Morison, called the paste 'Bipp' because it contained bismuth subnitrate, iodoform and paraffin.[104] Bismuth subnitrate, also known as bismuth hydroxide nitrate oxide and by other names, is prepared by the reaction of bismuth nitrate and water.

The following year another surgeon, also named Morison, described the use of magnesium sulfate cream to treat infected war wounds at the Red Cross and war hospitals in Sunderland.[105] The cream was prepared by mixing dry magnesium sulfate powder with a mixture of carbolic acid and glycerine.

An antiseptic powder prepared three years earlier at the University of Edinburgh, consisted of equal weights of chloride of lime and boric acid. It was called 'eupad'.[106] The powder dissolved in water to form hypochlorous acid: 'the most powerful antiseptic known', according to the Edinburgh team. The solution was named 'eusol', an acronym for 'Edinburgh University Solution of Lime'. It became one of the most widely employed antiseptic solutions during and after the First World War.

The use of an antiseptic to destroy pathogenic microorganisms relies essentially on chemical reactions between the antiseptic and the proteins and other constituents of the microorganism.[107] In a paper published in August 1915, English chemist Henry Drysdale Dakin (1880–1952) draws on the results of research carried out by French bacteriologist Maurice Daufresne that compared the ability of a range of antiseptics to destroy bacteria suspended in test tubes of water with their ability to destroy bacteria mixed with blood serum. The results clearly show that antiseptics are far less effective at killing bacteria in blood serum than in water. The germicidal power of carbolic acid, for example, is very low in blood serum and to be effective as a germicide has to be used at such a high concentration that it 'is destructive of healthy tissue'. Other antiseptics, such as mercuric chloride, are irritants even when used in dilute solution. Iodine, he observes, is useful for disinfecting skin but is less satisfactory for treating deep wounds.

Likewise, solutions of sodium hypochlorite which had high germicidal action was found to be irritating when applied to wounds. The irritation was caused by formation of the alkali sodium hydroxide when the hypochlorite dissolved in water. In his August 1915 paper, Dakin reports that a non-irritant antiseptic hypochlorite solution could be prepared by dissolving sodium carbonate and bleaching powder in water, filtering the solution to remove the calcium carbonate that was formed and then finally adding boric acid to the solution. The solution became known as 'Dakin's solution'.

The following year, Dakin teamed up with French surgeon and biologist Alexis Carrel (1873–1944) to carry out research on the treatment of infected wounds. In 1912, Carrel had won the Nobel Prize in Physiology or Medicine for pioneering new surgical techniques, most notably a method of using sutures, that is stitches, to close blood vessels. Working together at a temporary hospital in Compiègne, northern France, Carrel and Dakin developed the so-called Carrel-Dakin method of treating infected wounds. It involved flushing the whole surface of the wound at 'frequent intervals' with a 'gentle stream' of fresh Dakin's antiseptic solution.[108]

Dakin also continued work on the development of new antiseptic solutions. He was particularly interested in a family of organic compounds containing chlorine linked to nitrogen. These compounds, known as organic chloramines, possess strong germicidal properties. Dakin and co-workers published a paper in January 1916 reporting that an organic chloramine, which the authors called 'chloramine-T' to distinguish it from other chloramines, not only exhibited intense germicidal activity but was also 'practically' non-toxic and far less irritating than sodium hypochlorite.[109] It could therefore be used safely at higher concentrations and proved particularly useful in treating injuries to the jaw and mouth.

One of the chemical names of chloramine-T is sodium *p*-toluenesulfonchloramide. The compound, an organic chloramine sodium salt, was first prepared in 1905. The organic component contains a chlorine atom attached to a nitrogen atom. The salt readily dissolves in water but is practically insoluble in oils and many organic solvents, making it unsuitable for use as an antiseptic in applications such as creams and pastes.

In June 1917, Dakin and a fellow researcher showed that a related organic chloramine, also first prepared in 1905, not only possessed germicidal properties but was also soluble in a variety of oils and non-aqueous solvents.[110] The compound did not contain sodium and was not a salt. Each molecule of the organic compound included two chlorine atoms attached to a single nitrogen atom. Its chemical name is *p*-toluenesulfondichloramide. Dakin and his co-worker named it 'dichloramine-T'. Joshua Sweet, a surgeon in the United States Army Medical Reserve Corps, subsequently tested a solution of dichloramine-T in a chlorinated oil on the war wounds of some eighty soldiers at a British hospital in France, concluding that the solution 'is of great advantage in wound treatment'.[111]

In 1917 and 1918, a number of synthetic nitrogen-containing organic dyes were beginning to be studied and employed as antiseptics to treat war wounds. They included three dyes with similar chemical structures: brilliant green, malachite green and crystal violet – also known as Gentian violet. Creams, pastes and aqueous solutions of a deep orange antiseptic chemical known as acriflavine were also widely used in casualty clearing stations and military hospitals to disinfect wounds. Acriflavine, a mixture of two closely related nitrogen-containing organic compounds, was first synthesised in 1912 by Ehrlich, the German scientist who discovered salvarsan, the arsenical compound used to treat syphilis.

'Notoriously unsuccessful'

'In this war practically every wound is heavily infected', observed Sir Almroth Wright, the British pioneer of typhoid vaccinations and consultant physician to the BEF in France, in an address to the Royal Society of Medicine in London on 30 March 1915. He also pointed out that 'all manner of filth containing pathogenic organisms and spores' contaminated the clothes and skin of the soldiers, while projectiles carried these contaminants in and implanted them 'far beyond the reach of any prophylactic applications of antiseptics'.[112]

During the war, Wright carried out research on wound infections in a laboratory attached to a hospital in Boulogne-sur-Mer, France. He worked with Scottish bacteriologist Sir Alexander Fleming (1881–1955) who was to discover penicillin in 1928. In October 1915, Wright reported in another lecture at the Royal Society of Medicine that he had not encountered 'any satisfactory clinical or bacteriological evidence of the utility of antiseptics as employed in infected wounds'.[113]

Wright and Fleming showed that although antiseptics worked well on the surface of the skin, in a wound they only killed infectious microbes accessible on the surface of the wound and not those buried deep inside. At the same time, the antiseptics also tended to destroy the body's healthy cells that rush to a wound in order to scavenge and kill the invading microbes. In some cases, antiseptics even encouraged the multiplication of infectious microbes inside a wound and therefore promoted infectious disease and, as a consequence, death. Wright therefore advocated a surgical procedure that involved cleaning wounds with sterile hypertonic saline solutions rather than antiseptics, an approach that is widely practised today.

In November 1914, British surgeon Sir Anthony Bowlby had recommended washing wounds with an antiseptic such as hydrogen peroxide.[114] In his 1919 lecture, he reported that, by 1918, the use of strong antiseptics to treat gas gangrene had largely been abandoned.[34] He noted that the microorganisms that cause gas gangrene are resistant to even the strongest antiseptics and grow freely in damaged muscle, especially if a foreign body such as a piece of muddy clothing, rich with the organisms, is embedded in the wound. 'Timely excision and surgical cleansing of the wound and the removal of all foreign bodies' was therefore the best solution.

Maynard Smith, a consulting surgeon to the BEF, presented similar findings to American medical officers in a lecture in 1918.[115] He recommended washing and painting the skin around the wound with a dilute antiseptic solution of picric acid in ethanol. After removing debris, the surface of the wound should then be swabbed with eusol, while injecting antiseptics into wounds or syringing or irrigating the depths of wounds with antiseptics were futile and more likely to spread infection to previously uncontaminated areas. Strong chemical antiseptics were 'notoriously unsuccessful' and, whatever the treatment, severe wounds generally 'became infected and foul', with the wounded frequently dying or suffering a prolonged and painful convalescence.

10

Killing the Pain

In case of emergency

'There were necessarily many accidents' in the early days of flying owing to the 'structural weaknesses in the aeroplanes' and the inexperience of pilots, explains H. Graeme Anderson in an article on 'the medical aspects of aeroplane accidents' published in 1918.[1] Anderson, a London-based surgeon attached to the Royal Naval Air Service (RNAS), produced statistics to show that in one six-month period covering 4,000 hours' flying and 9,000 flights from one aerodrome where pilots trained, fifty-eight planes crashed and fifteen pilots were injured.

Anderson advises that in the event of an accident at an aerodrome, the look-out man at the sick bay or dressing station should proceed to the scene with a stretcher, a first-aid dressing bag and an emergency tool kit case consisting of two crowbars, two wire-cutters, a saw, a long stout knife, a hammer, cloth cutting scissors and a fire extinguisher. If the aviator was injured and pinned under the wreckage, he should not be dragged out – unless the plane is on fire – and rather 'the wrecked machine should be cut away from him'. The aerodrome duty surgeon should proceed to the scene by car or by foot, or 'if the accident is at a distance of a mile or more, it is better to go by aeroplane'. An ambulance with two sick-berth attendants should also be despatched to the scene, carrying a small bag containing a bottle of sterilised water, six first-aid dressings and slings, picric acid dressings, a tourniquet, cloth cutting scissors, a knife, a bottle of chloroform and face mask, brandy, two hypodermic syringes and morphine solution.

Morphine, or morphia as it was often called, was and still is the analgesic of choice for relieving severe and persistent pain. It was readily available during the early days of the war and some British officers at the time kept a supply with them in case of emergency. British author Vera Brittain's fiancé, Roland Leighton (1895–1915), was one such soldier. Before returning from leave to the Western Front in August 1915, he travelled with his fiancée to London where he bought a bouquet of deep red roses for her, a pipe for himself and also 'restocked his medical case with morphia'.[2] Brittain writes that she was pleased that he bought a good supply of the drug. At the end of December 1915 while fighting in France, Leighton was shot by a sniper and died of wounds at a casualty clearing station.

Morphine, like emetine and quinine which we met in the previous chapter, is an alkaloid. It works on the central nervous system to activate pain-relieving chemicals in the brain, spinal cord and gut. However, it has a number of adverse side-effects including blurred vision, confusion, loss of appetite, vomiting, respiratory depression, and severe constipation. Moreover, the alkaloid is highly addictive and requires higher and higher doses to maintain the same effect.

There are numerous references in medical reports to the use of the drug to relieve the suffering of soldiers, sailors and airmen wounded during the First World War. For example, in the Battle of Mons on 23 August 1914, the 80,000-strong BEF suffered enormous casualties. During the retreat from Mons, Arthur Osborne, a doctor in the RAMC, reports that he was overwhelmed with the number of wounded. He observes that all he could do was tend to the most badly injured by tightening their amateur tourniquets, plugging gaping chest wounds with gauze, and giving 'heroic doses' of morphia.[3]

As the alkaloid is only sparingly soluble in water, it was usually administered to the wounded either by mouth or by injection as a water-soluble salt such as morphine tartrate. In a report in August 1916, two RAMC captains refer to the use of the tartrate while treating infected gunshot wounds.[4] A quarter grain of the salt was usually given after dressing a wound 'as most patients complain of pain for a few hours'. A grain is a unit of mass equal to 64.8 milligrams (mg) or 0.0648 grams (g).

During the Battle of Jutland, fought on 31 May–1 June 1916, the British Royal Navy battlecruiser HMS *Tiger* was hit by a number of shells fired by battleships in the German Navy's High Seas Fleet. Many of the wounded on the ship received severe burns which in some cases proved fatal. The medical staff on the ship had established two stations, one forward and one aft, for receiving the wounded. The conditions in these stations were dire during the battle, as ventilation was poor and fumes from exploding shells and burning cordite posed a serious risk of suffocation even when respirators were worn. Stretcher parties brought the wounded to these stations, their wounds were dressed with gauzes and boric ointment was applied to burns.[5] The medical officers then injected each patient with a half grain dose of morphine and, if necessary, a similar dose thirty minutes later. In some cases, the officers placed morphine tablets under the tongues of the wounded, but this method of administering the drug proved unsatisfactory.

Wounded men on HMS *Lion*, the flagship of the British Royal Navy Grand Fleet, fared a little better. The ship's medicals officers gave each wounded man two thirds of a grain of morphine hypodermically: 'This large dose acted like a charm; pain was instantly relieved, and haemorrhage controlled'.

'The first true drug'

Morphine, a bitter white crystalline compound, is the principal alkaloid in the opium poppy *Papaver somniferum*. It can lay claim to being 'the original alkaloid and the first true drug', comments assistant professor Paul Blakemore and professor emeritus James White at Oregon State University in a feature article on the topic.[6] 'Early attempts to unlock the mysteries of opium provided a major stimulus to the development of organic chemistry'. It can even be argued, they say, that morphine spawned the entire field of medicinal chemistry. The powerful analgesic and euphoric properties of opium poppy pods have intrigued humans for thousands of years, although it was not until 1806 that pure morphine was first isolated from dried poppy resin. The alkaloid was named after Morpheus – the Greek god of dreams.[7]

The chemical structure of morphine was not established until 1925 and it would be another twenty-seven years before a chemical procedure for synthesising the alkaloid was published.[8] Chemical synthesis of this naturally-occurring compound, however, is neither practical nor commercially viable. The pure compound is produced commercially from opium resin or poppy straw using a series of extraction and purification steps involving water and organic solvents.

The commercial production of morphine and other alkaloids was spearheaded in the nineteenth century by the chemical and pharmaceutical company E. Merck, which had been founded by the Merck family in Darmstadt, Germany in 1668.[9] Merck & Co., almost 80 per cent of which was owned by E. Merck, was established in the United States in 1891 by George Merck (1867–1926). When the United States entered the First World War in 1917, George Merck had to sever his relationship with E. Merck and the German branch of his family. He voluntarily surrendered E. Merck's share of the American company to the United States Alien Property Custodian, and in 1919 the custodian put the Merck stock up of public action. With the help of two investment banks, Goldman Sachs and Lehmann Brothers, George Merck bought the stock, regained control of the company and re-established it as an independent company.

Therefrom or therewith

On 11 May 1916, the British Army Council issued an order listing a dozen intoxicant drugs that 'no person shall sell or supply … to or for any member of His Majesty's Forces unless ordered for him by a registered medical practitioner on a written prescription … marked "not to be repeated"'.[10] The order included analgesic drugs such as morphine, codeine, diamorphine, opium, and 'all other salts, preparations, derivatives, or admixtures prepared therefrom or therewith'.

The first two drugs, morphine and codeine, are two of the eighty or so alkaloids that occur naturally in the opium poppy. The chemical names of codeine include methylmorphine and morphine monomethyl ether. They show that codeine is a close chemical cousin of morphine, the only difference being that one of the hydrogen atoms in the morphine molecule is replaced by a methyl group, that is a carbon atom attached to there hydrogen atoms, in the codeine molecule. Although both chemicals are opiates, that is alkaloid drugs derived from the opium poppy, codeine is far less potent than morphine.

Likewise, diamorphine, also known as diacetylmorphine or more commonly as heroin, is chemically related to morphine. It was first synthesised in 1874 by the reaction of morphine and acetic anhydride, an organic compound derived from acetic acid. In 1897, diamorphine was prepared independently by a chemist at the German pharmaceutical firm Bayer. The following year the company put the product on the market under the proprietary name 'Heroin' as a cough medicine for children.

Diamorphine, like morphine, is a powerful narcotic analgesic that can become highly addictive if used repeatedly. It was sometimes employed along with morphine in the First World War. For example, two RAMC medical officers refer to the use of both analgesics in a tented casualty clearing station that received wounded men during activity on a sector of the Western front between 27 September and 15 October 1918.[11] The two officers describe resuscitation treatment in a ward located in a long marquee connected by a covered way to the station's operating theatre: 'Morphine judiciously given, heroin in chest cases, and camphor as a cardiac stimulant, were all utilized when desirable'.

Some drugs containing opium alkaloids were administered during the war not just as pain killers but also as sedatives. One such drug was omnopon, a yellowish-brown powder that readily dissolves in cold water. The powder contains salts of morphine, codeine and papaverine – another alkaloid extracted from the opium poppy. Omnopon was often given to 'restless' soldiers who had received gunshot wounds in the abdomen before they were taken to theatre for surgery.[12]

Opium was also commonly used as an analgesic and sedative. For instance, at the Battle of Arras in April 1917, British war poet and soldier Siegfried Sassoon complains in his memoir of suffering from gastritis, that is inflammation of the stomach, a sore throat and festering scratches on each hand.[13] To relieve the pain, one of his orderlies gave him an opium pill of 'constipating omnipotence'. The drug, which contains morphine, codeine and other alkaloids, is a sticky brown resin obtained by collecting and drying the latex that exudes from lanced poppy pods. The narcotic was popularised in sixteenth-century Europe when the Swiss physician and alchemist Paracelsus (1493–1541) introduced laudanum, a solution of opium in alcohol, also known as tincture of opium.

The tincture was employed for a variety of purposes to treat the sick and wounded during the First World War. It was used, for example, to alleviate 'the mental strain' of some of the victims of gas poisoning, note three RAMC medical officers in a report on '685 cases of poisoning by noxious gases used by the enemy' who were treated at a casualty clearing station in France in May 1915.[14] After administering the tincture, the officers observed, the patients quietened down and fell into a peaceful sleep.

The tincture was also used with starch as an enema to treat some of the side-effects of acute dysentery, reported one RAMC medical officer during a discussion on the subject at a meeting in Alexandria, Egypt, held on 17 October 1915.[15] The same officer also explained that Dover's powder, a mixture of opium and ipecacuanha, was 'given freely at night' to patients suffering from dysentery in order 'to produce sleep and to relieve nocturnal diarrhoea'. As we saw in the previous chapter, Dover's powder was also used to treat the symptoms of Spanish flu.

The 'ipecacuhana and opium powder' combination was among the drugs supplied, chiefly in tablet form, by the Germans to Alexander Macdonald, a temporary captain in the RAMC, after he was taken prisoner at an advanced dressing station in France in March 1918.[16] Following his capture and a 'tedious march spread over two days', he was sent to the prisoner section of a French hospital: 'We found it full to overflowing with British wounded who were urgently in need of attention, as they had received no further dressing than the first'. The Germans also provided him with anaesthetics, antiseptics, tetanus antitoxin, dressings, morphine, aspirin, veronal and a variety of other drugs. Veronal, also known as barbital or barbitone, is a long-acting barbiturate that was widely used as a sleep-inducing drug, a sedative, anaesthetic and analgesic. Barbitone was one of the dozen intoxicant drugs listed by the British Army Council in May 1916.[10]

General anaesthetics

According to Macdonald, there was no shortage of anaesthetics in the hospital where he served as a captive medical officer. The Germans supplied him with chloroform, ether and ethyl chloride, three of the principal anaesthetics used by medical officers on all sides throughout the war. All three anaesthetics are organic chemical compounds with relatively simple molecular structures. Chloroform and ether are volatile liquids, and ethyl chloride is a gas under normal conditions. The compounds were used as general anaesthetics during surgery and sometimes while wounds were being dressed. They were administered by inhalation of the vapours through a face mask and, for this reason, were sometimes called 'inhalation anaesthetics'.

Until the 1840s, surgery could be brutal. Patients undergoing surgery were often strapped to tables and had to endure the intense pain of the surgeon's knife. General anaesthesia, if practiced at all, was fairly crude, although opium or alcohol was sometimes administered to reduce the pain or render the patient unconscious.

General anaesthetics, whether administered by inhalation, injection or orally, induce unconsciousness temporarily by depressing the central nervous system. Local anaesthetics, on the other hand, dull or kill pain temporarily in a region of the body without loss of consciousness. For surgical operations, they are administered by injection and work by inhibiting nerve function around the site of injection.

Although general anaesthetics were readily available in many hospitals and casualty clearing stations during various phases of the First World War, there are accounts of shortages. For example, when British soldier Harry Patch was injured in the groin by a piece of shrapnel during the Battle of Passchendaele in 1917, he was taken to a dressing station where the stock of anaesthetic had already been used up on the badly wounded.[17] Four men were therefore employed to hold Patch down fully conscious while the doctor cut out the piece of metal.

Chloroform

Chloroform was possibly the first inhalation anaesthetic to be recognised by the medical profession and one of the mainstays of anaesthesia, especially early on in the First World War. The compound, discovered in 1831, was prepared by the addition of sulfuric acid to a mixture of the organic solvent acetone and bleaching powder, also known as calcium hypochlorite. It came into use as an anaesthetic for obstetrics in 1847. Six years later, Queen Victoria (1819–1901) summoned the English doctor Jon Snow (1813–1858), a pioneer in anaesthesia, to administer the anaesthetic during the birth of her eighth child, Prince Leopold, Duke of Albany (1853–1884). Snow had earlier invented an inhaler for anaesthetizing patients, consisting of a face mask connected by a flexible tube to a canister of liquid chloroform. This was known as the 'closed' method of administering the anaesthetic. For Queen Victoria, he used the so-called 'open-drop' or 'open 'method. With this method, the anaesthetist holds an open canister of the anaesthetic above a face mask with an underlying face pad and allows drops of the liquid to fall onto the mask.

British Army surgeons began to use chloroform while carrying out amputations on soldiers injured with gunshot wounds during the Crimean War (1853–1856).[18] There is little evidence that general anaesthetics were used at all before the war. For example, when the Royal Victoria Hospital on Southampton Water opened in 1856, it had 'no military anaesthetists and relied on chloroform alone for surgical procedures'.[19]

Soon after it came into use as an anaesthetic, reservations began to be expressed about its efficacy. It was well known at the time to be a potent and occasionally fatal depressant and there was always the worry that a patient would not wake from an anaesthetic-induced sleep. In 1854, British military surgeon Sir John Hall cautioned against the use of chloroform saying: 'The smart of the knife is a powerful stimulant; and it is much better to hear a man bawl lustily than to see him sink silently into the grave'.[18]

There were also other concerns about its use: when chloroform is exposed to light, it reacts with oxygen in the air to form the highly toxic gas phosgene. A solution to this problem was factored into the design of 'a portable motor operation theatre', described in the January 1915 issue of the *British Medical Journal*.[20] The designer, G.H. Colt, an assistant surgeon at the Aberdeen Royal Infirmary in Scotland, recommends that the theatre should be heated by superheated steam and not by electricity. Furthermore, 'it is important to shut off all naked lights from the operating room on account of the production of phosgene gas from the chloroform vapour, and on account of the risk of explosion when open ether is used'. Unlike chloroform, ether is highly flammable.

Ether

In chemistry, the term 'ether' applies to a class of organic chemicals with similar chemical structures. The one used as an inhalation anaesthetic in the First World War was diethyl ether, a volatile organic compound that has a long history. It was first synthesised in 1540 by the German botanist and pharmacist Valerius Cordus (1515–544). The compound is prepared by heating a mixture of ethanol and a dehydrating agent such as concentrated sulfuric acid. In the ensuing chemical reaction, two molecules of ethanol combine with the loss of one molecule of water. The diethyl ether distils off from the mixture.

Two Americans, the surgeon John Collins Warren (1778–1856) and dentist William Morton (1819–1868), publicly demonstrated the use of diethyl ether as a surgical inhalation anaesthetic at Massachusetts General Hospital, Boston, on 16 October 1846.[21] While Morton administered the anaesthetic, Warren removed a tumour from the jaw of the unconscious patient, Edward Gilbert Abbott (1822–1855), who then woke claiming that he had felt no pain. It was not the first time that ether had been used as an anaesthetic, as the previous month Morton had used it for a painless tooth extraction.

Diethyl ether was used in the First World War either by itself or in combination with another anaesthetic such as chloroform. 'The particular requirements of war surgery have given a great impetus to the use of ether as a general anaesthetic for routine work, and especially to its administration by the open method', comments anaesthetist Frederick Silk (1848–1942) in an article published in 1919.[22] Silk, who served as an RAMC medical officer during the war, had founded The Society of Anaesthetists in London in 1893. Based on his experience, he recommended that ether mixed with chloroform in the ratio 32:1 should be used as an anaesthetic.

Ethyl chloride

Between 1901 and 1915, another RAMC anaesthetist had employed an inhaler and ethyl chloride as a preliminary anaesthetic followed by ether to anaesthetise patients. However, his work during the war led him to conclude that 'mitigated ether', that is a mixture of ether and chloroform, was a better anaesthetic than ethyl chloride. From October 1915 onwards, he began to administer ether-chloroform mixtures in ratios ranging from 2:1 to 32:1 to wounded soldiers at a hospital in Birmingham. In a paper published in 1917, he revealed that there was 'no danger or trouble' with a 16:1 mixture apart from two cases 'in which respiration ceased'.[23] He also concluded that a 2:1 mixture of ether and chloroform was preferable to ethyl chloride as a preliminary anaesthetic. Not all anaesthetists agreed however. The following year Arthur Mills, an RAMC anaesthetist who had 'considerable experience in the administration of anaesthetics to soldiers' since the beginning of the war, concluded that a sequence that involved first administering ethyl chloride and then ether is 'the best method' for the induction of anaesthesia.[24]

Ethyl chloride was also used as an anaesthetic by the Germans. In his description of medical and surgical work as a prisoner of war, Macdonald notes that nursing sisters in many German hospitals administered ethyl chloride for certain dressings.[16] They raised the patient's arm and administered the anaesthetic until the arm dropped.

As its name indicates, ethyl chloride is closely related to ethyl alcohol, nowadays more commonly referred to as ethanol. It is a relatively simple organic compound that was first prepared in the fifteenth century by alchemist Basil Valentine. He made it by distilling hydrochloric acid, then known as muriatic acid, with ethanol, also known as 'spirit of wine'. He named the product 'muriatic ether'. The anaesthetic properties of the compound were first reported in 1847 by French physiologist and pioneer of anaesthesia Jean Pierre Flourens (1794–1867).[25]

Nitrous oxide

Chloroform, diethyl ether and ethyl chloride were not the only inhalation anaesthetics employed during the First World War. Nitrous oxide was also used, albeit infrequently as complicated apparatus was required to administer the gas. It was given either by itself, in a nitrous oxide-ether sequence, or in a mixture with oxygen.[26]

Nitrous oxide is an inorganic compound containing molecules made up of two nitrogen atoms and one oxygen atom. It is prepared by heating ammonium nitrate – the same compound that is used as an agricultural fertiliser and was a component of ammonal and other explosive mixtures used in the war. The colourless, sweet-swelling gas was discovered by English clergyman and chemist Joseph Priestley (1733–1804) in 1772. Priestley also isolated ammonia and a number of other gases the same year and is credited with the discovery of oxygen two years later.[27]

In 1799, Humphrey Davy, the English chemist who identified chlorine as an element, began to experiment with nitrous oxide. He demonstrated that when the gas is inhaled, it not only destroys physical pain but is also physically pleasurable and induces hilarity. For this reason, the compound soon became known as 'laughing gas'.

In a book on his research on nitrous oxide published in 1800, Davy describes using the gas to tackle toothache: 'I experienced an extensive inflammation of the gum, accompanied with great pain'.[28] After breathing three large doses of the gas, the pain diminished and 'the thrilling came on as usual'. Later in the book he suggests, that the gas 'may probably be used with great advantage during surgical operations'.[29]

However, it would be more than forty years before attempts were made to exploit nitrous oxide as an inhalation anaesthetic. In 1844, American dentist Horace Wells (1815–1848) allowed the gas to be used for a pain-free tooth extraction by another American dentist, John Riggs (1811–1885), although it would take another twenty years before the gas began to be widely used for tooth extractions.

Premedication

In February 1915, two surgeons described their experiences of surgery over the previous two months at the Anglo-American Hospital at Wimereux, France.[30] Most of the patients were suffering from bullet, shrapnel or shell wounds, the great majority of which 'were in a more or less septic condition'.

Almost all the surgery was carried out using chloroform as the anaesthetic. In a few cases, according to the surgeons, '1/4 grain of morphine, 1/100 grain of hyoscine and 1/100 grain of atropine' were injected into the patient before the operation. Such drugs, when administered to a patient before surgery, are known as 'premedication'.

Morphine premedication was used to alleviate pain. Hyoscine, also called scopolamine, is an alkaloid obtained from the nightshade family of plants and is a water-soluble viscous liquid that has a range of therapeutic properties, preventing nausea for example. Surgeons used it pre-operatively in the war to diminish glandular secretions during surgery. Atropine, another alkaloid with a variety of medical applications, was employed as premedication to dry up lung secretions that might be inhaled during anaesthesia.

Local anaesthetics

In their report on surgery at Wimereux, the two surgeons add that in the one or two cases of extreme shock they encountered, they employed spinal anaesthesia rather than chloroform as a general anaesthetic. In medicine, the term 'shock' is used for a condition in which blood pressure suddenly drops to such an extent that it is unable to maintain an adequate circulation of blood around the body and therefore an adequate supply of oxygen to tissues. The condition is also known as acute circulatory failure.

Spinal anaesthesia is a form of anaesthesia in which a local anaesthetic is injected into the space around the spinal cord. It is used when general anaesthesia is unsuitable. During spinal anaesthesia, the patient remains conscious during the operation. For this reason, the two surgeons at Wimereux administered not only the spinal anaesthetic, but also a 'very small quantity' of ether by the open method to help 'overcome the mental shock which the amputation of a limb produces on a patient who is fully conscious'.[30]

The local anaesthetic they injected into the spine was a solution of a crystalline nitrogen-containing organic compound with the chemical name amylocaine hydrochloride. The compound was first prepared in 1904 by the French pharmaceutical chemist Ernest Fourneau (1872–1949). *Fourneau* is the French word for 'stove' and the chemist therefore named the compound 'stovocaine'.

The first local anaesthetic to be used in surgery was not a synthetic compound like stovocaine but rather a natural product, namely cocaine. The compound, an alkaloid, was isolated from the leaves of the coca plant and first used as a local anaesthetic in surgery in 1884.[31] Cocaine was one of the restricted drugs listed by the British Army Council in 1916 as it is habit-forming and has several other adverse side-effects. In the years leading up to the First World War, a number of synthetic nitrogen-containing organic compounds without the drawbacks of cocaine began to emerge as local anaesthetics. They included beta-eucaine, a compound with a complicated chemical structure similar to that of cocaine, albeit not as addictive.

One of the most widely used local anaesthetics during the war was another synthetic nitrogen-containing organic compound known as procaine or novocaine. The compound was first synthesised in 1905 by German chemist Alfred Einhorn (1856–1917) and has a chemical structure similar to that of cocaine. However, although it is less potent as an anaesthetic than cocaine, it is far safer to the extent that it is not addictive and is much less toxic.

Shortages

Before the beginning of the war, Britain, France and their allies purchased considerable quantities of synthetic drugs, like aspirin and novocaine, and naturally-occurring alkaloid drugs from Germany. When the war started, German supplies of these drugs to the Allies as well as supplies of dyes and other chemicals rapidly dried up.

In Britain, companies such as Burroughs Wellcome and May & Baker manufactured a limited range of drugs, but only in relatively small quantities. In some cases they had relied on German companies to grant licenses for their manufacture before the war and for the supplies of the chemical intermediates needed for their synthesis. Furthermore, at the start of the war, Britain was desperately short of industrial chemists and chemical engineers, the industrial know-how, and the industrial capacity required to manufacture the wide range of drugs in the quantities needed for the war effort. To tackle these problems, the government embarked on a crash programme of drug production. First of all, it abrogated German patents for the syntheses of drugs such as aspirin and then encouraged existing chemical manufacturing companies to enlarge their production plants.

As part of the programme, the British government also enlisted the help of university and technical college chemistry departments, not only to develop chemical processes for making the drugs but also to manufacture them. 'As late in the war as 1917, the universities were still supplying chemicals for the war effort' note Marlene and Geoffrey Rayner-Canham, experts in the history of chemistry.[32] The chemicals included alkaloids like atropine and local anaesthetics such as beta-eucaine and novocaine needed for military hospitals.

A project to make beta-eucaine, for instance, involved seventeen laboratories from around the country.[33] The collaboration led to an improved method of synthesis and the production of around 100kg of the anaesthetic, equivalent to more than 3 million doses.

Women such as Martha Whiteley (1865–1956), a chemistry lecturer at Imperial College London, played a key role in this endeavour. She assembled a team of female chemists to produce not only beta-eucaine and other drugs, but also chemical warfare agents.[34] In the years leading up to the war, there was a significant increase in the numbers of women taking chemistry degrees, the Rayner-Canhams observe. Without this increase 'the British war machine would have faced a severe shortage of chemists'.[32] As a consequence, the country would have inevitably faced a severe shortage of the chemicals need to sustain the war effort.

11

A Double-Edged Sword

War chemistry

The industrial-scale carnage and destruction of the First World War would not have been possible without the industrial-scale production of a vast variety of chemicals. The construction and operation of weapons, battleships, fighter aircraft, airships, submarines, and the merchant vessels and railways that carried troops, horses, supplies, and munitions to the various fronts of the First World War, all relied on the production of chemical materials.

The applied chemistry of the war, what might also be called 'war chemistry', underpinned the military technology of the war, drawing on almost every aspect of mainstream chemistry to achieve the military aims of the war. It exploited elements, such as hydrogen for airships, chlorine for chemical warfare and phosphorus for incendiaries. It produced inorganic compounds, such as the nitric acid needed to make explosives, and it developed more efficient processes for their manufacture. And for the first time in warfare, it applied knowledge of organic synthesis to convert organic chemicals extracted from coal tar and other sources to militarily-useful organic compounds, most notably the explosive trinitrotoluene.

The chemistry of the First World War demonstrated its destructive power on three major fronts. First of all, it generated the huge quantities of explosives required for artillery shells and fuses, for pistol, rifle and machine-gun cartridges, for grenades and trench mortar bombs, and for the mines blown up in tunnelling operations. It also created the chemical warfare agents that filled artillery gas shells or were released in cloud gas operations. And finally, war chemistry helped to provide the metals and alloys used to manufacture the guns and ammunition.

If chemistry was central to the war effort, then who were the war chemists? Academic, government and industrial chemists all played their part. In Britain, they worked with chemical engineers and their fellow scientists and technologists 'for King and country' at the behest of the military and political leaders who, along with the chemical manufacturers, encouraged, demanded and facilitated the development of chemical processes and the manufacture of chemicals for the war effort.

But much of the chemistry on which the war was based did not evolve because of the war – it evolved before the war. Most of the chemists in the hundred or so years leading up to the war who had made key discoveries did so out of sheer curiosity and in all innocence without any thought of potential military applications. They were not chemists in military service, but mostly civilian chemists. The job of the war chemists was to mobilise mainstream chemistry and bring it into play for military purposes. They examined previously discovered chemical processes and chemicals, they improved the processes and studied the properties of the chemicals, they selected the most suitable chemical processes and chemical products for the war effort, and they developed analytical techniques for establishing the identities, compositions and purity of these war products.

Whereas warfare has spawned numerous technological inventions, the Mills bomb and the machine-gun are just two examples, war has rarely if ever been the mother of invention in chemistry. For example, the mixture of chemicals known as gunpowder or black powder was discovered well before the invention of the gun. As far as chemistry is concerned, invention was the mother of the First World War. Chemistry, largely based on discoveries and inventions of the previous hundred years, mothered the war to the extent that it nurtured it with war materials: explosives, chemical warfare agents and metals.

The protective role

'Science has perfected the art of killing – why not of saving?' asks French surgeon and Nobel Prize winner Alexis Carrel in an article published in 1915.[1] As we saw in the previous chapter, Carrel collaborated with English chemist Henry Dakin to develop the Carrel-Dakin method of treating infected wounds with antiseptic solution. Two years later, Dakin points out in an article on 'biochemistry and war problems' that the destructive character of warfare would not be possible without the cooperation of the chemist, the physicist and the engineer.[2] He then adds that 'a large amount of purely scientific endeavour has been devoted to the conservation of the health of those taking an active part in the present struggle'.

As we saw in the two previous chapters, chemistry played a key role in fighting infection during the First World War and alleviating the pain and suffering of sick and wounded troops. Disinfectants, antiseptics, analgesics, anaesthetics and other pharmaceutical products are all chemicals produced using chemical processes devised by chemists and chemical engineers. The application of chemistry to the health and well-being of the armed forces, to hygiene and sanitation in the trenches, casualty clearing stations, and military hospitals, and to the care of the sick and wounded may be regarded as another aspect of war chemistry.

However, chemistry did not just save and conserve life with its array of contributions to medicine, surgery and anaesthesia. It also contributed to extending the lives of the warriors – troops, sailors and airmen – in other ways. Chemistry, notably electrochemistry, helped to provide the metals and alloys for the manufacture of steel helmets and armour for tanks and battleships. It provided the metals for the corrugated iron and barbed wire that protected trenches and dugouts, and for the tin cans that contained food. Chemistry also provided the drab mineral and synthetic coal tar dyes necessary for camouflage – for example, for the khaki or field grey uniforms that enabled soldiers to melt into the background and thus avoid enemy rifle fire.

The double edge

The war chemistry of the First World War was a double-edged sword. On one hand, it had the power to debilitate, damage and destroy, while on the other hand it had the capacity to protect, conserve and prolong life. It was used for offence and also for defence: for example, it was applied to generate toxic gases for attacking the enemy, and at the same time to create gas mask chemicals that provided protection against these deadly gases.

In some instances, the same chemistry and same chemicals were used for killing and protection, chlorine being a prime example. The element was employed in chemical warfare both as a weapon of mass destruction and to synthesise other chlorine-containing chemical warfare agents such as phosgene and mustard gas. It was also used either by itself or in the form of hypochlorous acid for water purification and disinfection and therefore for disease prevention.

The explosive properties of the inorganic chemical ammonium nitrate were exploited in blasting explosives such as ammonal. Yet ammonium nitrate is also an important fertiliser that boosts agriculture productivity and therefore helps to keep people alive. Picric acid is another example: the yellow organic chemical was used not only as a high explosive during the war, but also as a dye, as an antiseptic and to treat burns. Nitroglycerine, another organic chemical, was used both as an explosive and as a drug during the war.

Paradoxically, the protective edge of chemistry had a downside. By helping to reduce the human wastage of the war through the treatment of the sick and the care of the wounded, it enabled increasing numbers of men to return to the battlefields where they were once again exposed to the destructive edge of chemistry. It facilitated the more efficient use of manpower and, as a consequence, further death and destruction.

How chemistry changed the war

Chemistry changed the First World War; the war did not change chemistry, at least not to any significant extent. Chemistry was instrumental in determining the nature, shape and duration of the war. Its application resulted in death in an unprecedented manner and on an unimaginable scale: the artillery bombardments, and the trench and tunnel warfare would not have been possible without chemicals of one sort or another; every round of ammunition, whether for a pistol, rifle, machine-gun, trench mortar, howitzer, or siege gun relied on the manufacture of explosives and metals.

Furthermore, the war would have ended within a year had not chemists and their colleagues previously found ways of manufacturing explosives from nitrogen in the air and nitrate minerals, and the chemicals extracted from coal tar and other sources.

The so-called chemists' war of tear gas, choking gas, nerve agents, blistering agents, smokes and incendiaries would not have been possible without the inventions and developments in chemistry of the previous century. Similarly, the chemistry and chemicals that led to improvements in hygiene, sanitation, and the care of the sick and wounded also prolonged the war by enabling increased numbers of fighting men to continue fighting.

Chemicals are intrinsically neither good nor bad: they are inanimate and therefore have no moral or ethical value in themselves. It is the use to which they are put and how they are used that may be considered good or bad. We, as human beings, can either apply chemistry for constructive, protective and other beneficial purposes, or for destructive reasons.

Chemistry, more than any other branch of science, 'has provided a cornucopia of good things, both of necessities and luxuries, which have improved our health, our wealth, and also I believe our happiness' said Lord George Porter (1920–2002) in a plenary lecture at the 30th Congress of the International Union of Pure & Applied Chemistry held in Manchester in 1985.[3] Porter, who won the Nobel Prize in Chemistry in 1967, gave numerous examples of chemistry's cornucopia. He pointed out, for example, that during the twentieth century our health improved out of all recognition and our average lifespan doubled through better nutrition, better hygiene and the use of an ever-increasing array of pharmaceutical products. Yet chemistry's cornucopia, which we all enjoy, has a number of downsides, including pollution, depletion of non-renewable resources and the wastage of materials. And in the First World War a multiplicity of chemicals was used to kill, maim and wound humans, and to damage and destroy buildings, farms, roads and railways alike.

The war demonstrated chemistry's ability to act as a double-edged sword in a way that had never been seen before. The inventions and discoveries of chemists in the hundred or so years before the start of the war provided a cornucopia of good things for the twentieth century, but at the same time opened up a Pandora's box of death and destruction.

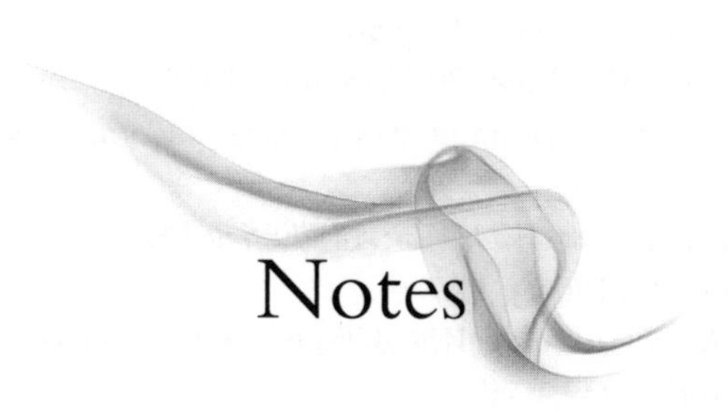

Notes

Where a short note is provided, more details are given in the bibliography on p. 233–6.

1. The Chemists' War

1. Baekeland, L.H., 'The Naval Consulting Board of the United States', *Ind. Eng. Chem.*, 1916, 8, p.67.
2. Pilcher, R.B. 'Chemistry in wartime', *Ind. Eng. Chem.*, 1917, 9, p.4.
3. Rhees, D.J., 'The chemists' war: the impact of World War I on the American chemical profession', *Bull. Hist. Chem.,* 1992–1993, 13–14, p.40–7.
4. MacLeod, R., 'The chemists go to war: the mobilisation of civilian chemists and the British war effort, 1914–1918', *Annals of Science*, 1993, 50, p.455–81.
5. Sanderson, M., 'The University of London and industrial progress 1880–1914', *Journal of Contemporary History*, 1972, 7, p.243–62.
6. Whitney, W.R., 'England's tardy recognition of applied science', *Ind. Eng. Chem.*, 1915, 7, p.819–20.
7. Munroe, C.E. A, 'lesson from history – lest we forget', *Ind. Eng. Chem.*, 1922, 14, p.75.
8. LeMaistre, F. J., 'Conditions of the French chemical industries during 1916', *Ind. Eng. Chem.*, 1918, 10, p.421–23.
9. Fink, C.G., 'Greetings from the electrochemists', *Ind. Eng. Chem.*, 1917, 9, p.1006.
10. Culbertson, W.S., 'The Tariff Commission and our chemical industries', *Ind. Eng. Chem.*, 1917, 9, p.1013-1016.
11. Tone, F.J., 'The exposition in war and in peace', *Ind. Eng. Chem.*, 1918, 10, p.828-29.

12. Bailey, G., 'Modern project management and the lessons from the study of the transformation of the British Expeditionary Force in the Great War', *Management Decision*, 2005, 43, p.56-71.
13. Anon., *The New York Times*, 20 November 1915.
14. Howe, H.E., 'Chemical industry in Canada', *Ind. Eng. Chem.*, 1917, 9, p.548–51.
15. Taylor, A.J.P., *The First World War: An Illustrated History*, Penguin Books, Harmondsworth, 1966, p.11.
16. Prentiss, A.M., *Chemicals in War: A Treatise on Chemical Warfare*, McGraw-Hill, New York, 1937, p.vii.
17. Earle, R., 'Chemistry and the Navy', *Ind. Eng. Chem.*, 1919, 11, p.924-27.
18. Hellemans, A., Bunch, B., *The Timetables of Science: A Chronology of the Most Important People and Events in the History of Science*, Simon and Schuster, New York, 1988, p.426–33.
19. Editorial, 'On our opportunities', *Ind. Eng. Chem.*, 1914, 6, p.794.
20. Nichols, W.H., 'The war and the chemical industry', *Ind. Eng. Chem.*, 1915, 7, p.131–32.
21. Gudeman, G., 'Aspects of some chemical industries, in the United States, today', *Ind. Eng. Chem.*, 1915, 7, p.151–55.
22. Editorial, 'The modern miracle', *Ind. Eng. Chem.*, 1918, 10, p.508.
23. Baekeland, L.H., 'The future of chemical industry in the United States', *Ind. Eng. Chem.*, 1917, 9, p.1020.
24. Jones, G., 'War disturbances and peace readjustments in the chemical industries', *Ind. Eng. Chem.*, 1918, 10, p.783–85.
25. Kies, W.S., 'The development of our export trade', *Ind. Eng. Chem.*, 1917, 9, p.1018-19.

2. Shell Chemistry

1. Westwell, p.122.
2. Stevenson, p.103.
3. Jünger, p.27.
4. Graves, p.116.
5. Hogg, p.6.
6. War Office, p.140.
7. War Office, p.47.
8. Prentiss, p.434.
9. Prentiss, p.436.
10. Prentiss, p.442.
11. Prentiss, p.443.
12. Prentiss, p.161.

13. Prentiss, p.455.
14. Prentiss, p.463.
15. Emsley, p.133.
16. War Office, p.39.
17. Jünger, p.166.
18. Macdonald, L., p.172.

3. Mills Bombs and other Grenades

1. Ainslie, p.9.
2. Sassoon, p.65.
3. Richards, p.93.
4. Obituary: Sir William Mills, *The Times*, 8 January 1932, p.12.
5. Hartcup, p.62.
6. Saunders, p.161.
7. Croddy, p.118.
8. Prentiss, p.297.
9. Jones, p.45.

4. The Highs and Lows of Explosives

1. McEwen, p.98.
2. Haythornthwaite, p.88.
3. War Office, p.218.
4. War Office, p.178.
5. Clarke, p.36.
6. Clarke, p.35.
7. Cocroft, W.D., 'First World War explosives manufacture: the British experience', in MacLeod, p.31.
8. Cocroft, W.D., 'First World War explosives manufacture: the British experience', in MacLeod, p.36.
9. Munro, C. E., 'Contributions of the chemist to the explosives industry', *Ind. Eng. Chem.*, 1915, 7, p.945.
10. MacLeod, R.M., 'Chemistry for King and Kaiser', in Travis, p.39.
11. Hopkins, O.P., 'Effect of war on American chemical trade', *Ind. Eng. Chem.*, 1918, 10, p.692–700.
12. Hopkins, O.P., 'Our foreign trade in chemicals', *Ind. Eng. Chem.*, 1920, 12, p.840–44.
13. Hartcup, p.55.
14. Brown, p.172.

15. Storey, p.37.
16. Storey, p.38.
17. Anon., 'Health of Munition Workers', *Br. Med. J.*, 1916, 1, p.488.
18. Anon., 'Trinitrotoluene Poisoning', *Br. Med. J.*, 1916, 2, p.842.
19. Thursfield, H., 'Note upon a case of jaundice from trinitrotoluol poisoning', *Br. Med. J.*, 1916, 2, p.619.
20. Warner, p.116.
21. Faulks, p.372-3.
22. Akhavan, p.1.
23. Russell, p.2.
24. Macksey, p.132.
25. War Office, p.4.
26. Jünger, p.93.
27. War Office, p.35.
28. Holt, p.72.
29. Woodbridge, R.G. Jr, 'Nitrocellulose from wood pulp', *Ind. Eng. Chem.*, 1920, 12, p.380–84.
30. Marsh, L.G., 'Possible uses of corncob cellulose in the explosives industry', *Ind. Eng. Chem.*, 1921, 13, p.296-98.
31. Barton, p.128.
32. Akhavan, p.2.
33. O'Neil, p.799.
34. Ainslie, p.53.
35. Laszlo, P., 'George Darzens (1867–1954): inventor and iconoclast', *Bull. Hist. Chem.*, 1994, 15/16, p.59–64.
36. Stevens, K.K., 'The inspection and testing of trinitrotoluene', *Ind. Eng. Chem.*, 1917, 9, p.801–3.
37. Barton, p.118.
38. Marshall, p.14.
39. Akhavan, p.4.
40. Hartcup, p.49.
41. Ainslie, p.51.
42. Hamilton, p.40.
43. Davis, p.368.
44. Howard, E., *Phil. Trans. Roy. Soc.*, 1800, p.204.
45. Wooton, P., 'Washington letter', *Ind. Eng. Chem.*, 1917, 9, p.990.
46. Ray, A.B., 'Incendiaries in modern warfare', *Ind. Eng. Chem.*, 1921, 13, p.714.
47. Prentiss, p.366, 469.

5. The Metals of War

1. Scott, S., 'British control of platinum', *Ind. Eng. Chem.*, 1917, 9, p.731.
2. McDonald, p.229.
3. McMillan, A., 'Sulfuric acid manufacture', *Ind. Eng. Chem.*, 1920, 12, p.181.
4. Roebuck, P., 'A great British invention', *Chemistry in Britain*, July 1996, p.38.
5. Freemantle and Tidy, p.86.
6. McConnell, R.E., 'The production of nitrogenous compounds synthetically in the United States and Germany', *Ind. Eng. Chem.*, 1919, 11, p.837.
7. Hartley, H., 'Military importance of the German chemical industry', *Ind. Eng. Chem.*, 1921, 13, p.284.
8. Crowther, p.96.
9. McDonald, p.390.
10. McDonald, p.392.
11. White, A.H., 'The present status of nitrogen fixation', *Ind. Eng. Chem.*, 1919, 11, p.231.
12. McDonald, p.391.
13. Merz, A.R., 'Russia's production of platinum', *Ind. Eng. Chem.*, 1918, 10, p.920.
14. McDonald, p.306.
15. Anon., 'An appeal to the wives and daughters of chemists', *Ind. Eng. Chem.*, 1917, 9, p.445.
16. Anon., 'The platinum situation', *Ind. Eng. Chem.*, 1917, 9, p.544.
17. Parsons, C.L, 'Platinum in jewelry', *Ind. Eng. Chem.*, 1917, 9, p.622.
18. Anon., 'Platinum scraps', *Ind. Eng. Chem.*, 1918, 10, p.336.
19. Anon., 'Platinum oscillations', *Ind. Eng. Chem.*, 1918, 10, p.95.
20. Anon., *The New York Times*, 1 March, 1918.
21. Anon., 'Man of steel: Henry Bessemer', tce *(the chemical engineer)*, December 2010/January 2011, 834, p.27.
22. Sargent, G.W., 'Contributions of the chemist to the steel industry', *Ind. Eng. Chem.*, 1915, 7, p.932.
23. Cushman, A.S., 'Contributions of the chemist to the iron and steel industry', *Ind.Eng. Chem.*, 1915, 7, p.934.
24. Travis, T., 'The Haber-Bosch process: exemplar of 20th century chemical industry', *Chemistry & Industry*, 2 August 1993, p.581.
25. Forty, p.30.
26. Anon., 'Armour', *Br. Med. J.*, 1916, 2886, p.603.
27. Beistle, C.P., 'Shipping containers', *Ind. Eng. Chem.*, 1919, 11, p.674.
28. Singmaster, J.A., 'Reconstruction in the zinc industry', *Ind. Eng. Chem.*, 1919, 11, p.146.

29. Anon., 'German utilization of iron-furnace slag', *Ind. Eng. Chem.*, 1914, 6, p.261.
30. Olsen, K.K., 'Three hundred years of assaying American iron and iron ores', *Bull. Hist. Chem.*, 1995, 17/18, p.41.
31. Anon., 'British metals and munitions', *Ind. Eng. Chem.*, 1916, 8, p.190.
32. Botting, p.36.
33. War Office, p.282.
34. Eaton, p.40.
35. War Office, p.207.
36. Anon., 'The approximate melting point of some commercial copper alloys', *Ind. Eng. Chem.*, 1914, 6, p.164.
37. McAuliffe, p.23.
38. Solly, E., 'Report on the death of Henry Hennell', 1842, *Notes and Queries*, 1877, s5-VII(183), p.505.
39. War Office, p.194.
40. Crowther, p.132.
41. Russell, p.85.
42. War Office, p.45.
43. War Office, Plate LI and p.309.
44. Anon., 'German views on British bullets', *Br. Med. J.*, 1915, 1, p.1023.
45. Anon., 'German, French, and British bullets', *Br. Med. J.*, 1914, 2, p.990.
46. Bland-Sutton, J., 'Value of radiography in the diagnosis of bullet wounds', *Br. Med. J.*, 1914, 2, p.953.
47. Haythornthwaite, p.63.
48. Anon., 'German and British bullets', *Br. Med. J.*, 1914, 2, p.1041.
49. Anon., 'British and German small arm ammunition', *Br. Med. J.*, 1914, 2, p.895.
50. Anon., 'Home hospitals and the war', *Br. Med. J.*, 1915, 1, p.86.
51. War Office, p.174.
52. Maltby Clague, T. and Watson, A.J., 'Experiments on the solubility of plain lead and antimonial lead', *Br. Med. J.*, 1917, 2, p.757.
53. Hanson, p.47.
54. British Fire Prevention Committee, Air raids, 'Precautions against incendiary bombs', *Br. Med. J.*, 1915, 1, p.933.
55. Hartcup, p.159.
56. McMillan, A., 'The Buckingham bullet', *Ind. Eng. Chem.*, 1919, 11, p.883.
57. Abbott, p.22.
58. The Airship Heritage Trust, http://www.airshipsonline.com/airships/ss/index.html, 18 July 2011.
59. Freemantle, p.444.
60. Remarque, p.1.
61. Wakefield, p.99.
62. Wakefield, p.157.

63. Hall, p.86.
64. Farmborough, p.254.
65. Holmes, p.148.
66. Partington, p.490.
67. Adams, p.65.
68. Graham, J.C., 'The French connection in the early history of canning', *Journal of the Royal Society of Medicine*, 1981, 74, p.374–81.
69. Remarque, p.109.
70. Arthur, p.116.
71. Saunders, p.9.
72. War Office, p.270, 574.
73. Haus, A.G., and Cullinan, J.E., 'Screen film processing systems for medical radiography: a historical review', *RadioGraphics*, 1989, 9, p.1203.
74. Badger, p.135.
75. Haythornthwaite, p.368.
76. Macdonald, G., p.80.
77. Macdonald, G., p. 82.
78. Haythornthwaite, p.369.
79. Hoare, p.227.
80. *War Neuroses: Netley Hospital, 1917* [electronic resource]: Welcome Film, 2008.
81. Willmott, p.189.
82. Anon., 'The new photography', *Br. Med. J.*, 1896, 1, p.289.
83. Davidson, J.M., 'Discussion on the localisation of foreign bodies (including bullets, &c.) by means of X-rays', *Proc. R. Soc. Med.*, 1915; 8 (Electro Ther Sect); p.1–8.
84. Barclay, A.E., 'Discussion on new methods for localisation of foreign bodies', *Ibid.*, p.19–21.
85. Griffiths, B., 'Elements of inspiration', *Chemistry World*, January 2011, p.42; Curie, P. and Sklodowska-Curie, *'Sur une substance nouvelle radio-active, contenue dans la pechblende'*, *Compt. Rend. Acad. Sci.*, 1898, 127, p.175.
86. Carlino, S., 'Marie Curie and the centennial elements', *Education in Chemistry*, November 1998, p.151.
87. Pasachoff, N., 'Marie Curie', *Chem. & Eng. News*, 27 June, 2011, p.66.
88. Anon, 'Obituary: Florence A. Stoney, OBE, MD', *Br. Med. J.*, 1932, 2, p.734.
89. Leneman, L., 'Medical women at war, 1914–1918', *Medical History*, 1994, 38, p.160.
90. Ramsay, M.L. and Stoney, F.A., 'Anglo-French Hospital, No. 2, Château Tourlaville, Cherbourg', *Br. Med. J.*, 1915, 1, p.966.
91. Thurstan Holland, C., 'Radiology in clinical medicine and surgery', *Br. Med. J.*, 1917, 1, p.285.

6. Gas! GAS! Quick, boys!

1. Brittain, p.375.
2. Hitler, p.162.
3. Lewis, p.161.
4. Evans, p.14.
5. Croddy, p.6, 88.
6. Prentiss, p.4.
7. Fries, A.A., 'Chemical warfare', *Ind. Eng. Chem.*, 1920, 12, p.423.
8. Anon, *Science*, 323, 30 January 2009, p.565.
9. Hoare, p.150.
10. Croddy, p.131.
11. Prentiss, p.6.
12. Prentiss, p.661.
13. Ryan, p.39.
14. Croddy, p.146.
15. Carr, F.H., 'Gas warfare: the Harrison Memorial Lecture', *Br. Med. J.*, 1919, 2, p.140.
16. Prentiss, p.206.
17. Prentiss, p.679.
18. Hartcup, p.116.
19. Harris, p.12.
20. Corrigan, p.173.
21. Jones, p.58.
22. Palazzo, p.2.
23. Ryan, p.40.
24. MacLeod, R.M., 'Chemistry for King and Kaiser', in Travis, p.26.
25. Lee, p.148.
26. Prentiss, p.653.
27. Corey, J.H., Smart, J.K. and Hill, B.A., 'History of chemical war', in Tuorinsky, p.37.
28. Hartcup, p.94.
29. Prentiss, p.139.
30. Perkin, W.H. and Duppa, B.F., '*Ueber die Einwirkung des Broms auf Essigsäure*', *Justus Liebigs Annalen der Chemie,* 1858, p.106, 108.
31. Croddy, p.118.
32. Haber, p.20.
33. Prentiss, p.130.
34. Hartcup, p.95.
35. Perkin, W.H. and Duppa, B.F., '*Ueber die Jodessigsäure*', *Justus Liebigs Annalen der Chemie,* 1859, p.112, 125.
36. Prentiss, p.207.

37. Charles, p.157.
38. Prentiss, p.663.
39. Davy, J., 'On a gaseous compound of carbonic oxide and chlorine', *Philosphical Transactions of the Royal Society of London*, 1812, 102, p.144.
40. Ryan, p.13.
41. Coffey, p.103.
42. Ryan, p.75.
43. Prentiss, p.155.
44. Jones, p.20.
45. Nye, p.188.
46. Ryan, p.167.
47. Croddy, p.96.
48. Ryan, p.23.
49. Prentiss, p.159.
50. Norris, J.F., 'The manufacture of war gases in Germany', *Ind. Eng. Chem.*, 1919, 11, p.817.
51. Stenhouse, J., 'On chloropicrine', *Philosophical Magazine Series 3*, 1848, 33, p.53.
52. Anon., Anniversary meeting, *J. Chem. Soc., Trans.,* 1881, 39, p.185.
53. Brooks, N.M., 'Munitions, the military, and chemistry in Russia', in MacLeod, p.75.
54. Hartcup, p.104.
55. Goss, B.C., 'An artillery gas attack', *Ind. Eng. Chem.*, 1919, 11, p.829.
56. Zanetti, J. E., 'Interallied organizations for chemical warfare', *Ind. Eng. Chem.*, 1919, 11, p.721.
57. Jones, p.41.
58. Evans, p.40.
59. Hartcup, p.109.
60. O'Neil, p.193.
61. Haber, p.189.
62. Nye, p.189.
63. Hartcup, p.110.
64. Burrell, G.A., 'The Research Division, Chemical Warfare Service', U.S.A., *Ind. Eng. Chem.*, 1919, 11, p.93.
65. Coffey, p.109.
66. Croddy, p.102.
67. Harris, p.27.
68. Charles, p.170.
69. Prentiss, p.662.
70. Corey, J.H., Smart, J.K. and Hill, B.A., 'History of chemical war', in Tuorinsky, p.31.
71. Prentiss, p.190.

72. Prentiss, p.195.
73. Vilensky, Title.
74. Vilensky, p.4.
75. Harris, p.3; Lee, p.77.
76. Blatchley III, E.R. and Cheng, M., 'Reaction mechanism for chlorination of urea', *Environ Sci. Technol.*, 2010, 44, p.8529; De Laat, J., Feng, W., Freyfer, D.A., Dossier-Berne, F., 'Concentration levels of urea in swimming pool water and reactivity of chlorine with urea', *Water Res.*, 2011, 45, p.1139.
77. Auld, Chap.2.
78. Auld, S.J.M., 'Methods of gas warfare', *Ind. Eng. Chem.*, 1918, 10, p.297.
79. Bull, S., 'The early years of the war', in Sheffield, p.207.
80. Haber, p.70.
81. Hartcup, p.103.
82. Haber, p.76.
83. Lamb, A.B., Wilson, R.E. and Chaney, N.K., 'Gas mask absorbents', *Ind. Eng. Chem.*, 1919, 11, p.420.
84. Brooks, N.M., 'Munitions, the military, and chemistry in Russia', in MacLeod, p.88.
85. Dewey, B., 'Production of gas defense equipment for the army', *Ind. Eng. Chem.*, 1919, 11, p.185.
86. Coffey, p.113.
87. Fieldner, A.C., Teague, M.C. and Yoe, J.H., 'Protection afforded by army gas masks against various industrial gases', *Ind. Eng. Chem.*, 1919, 11, p.621.
88. Prentiss, p.567.
89. Haber, p.106.
90. Evans, p.44.
91. Haber, p.107.
92. Bret, P., 'Managing chemical expertise: the laboratories of the French Artillery and the Service des Poudres', in MacLeod, p.203.
93. Brooks, N.M., 'Munitions, the military, and chemistry in Russia', in MacLeod, p.86.
94. Haber, p.108.
95. Harris, p.21.
96. Evans, p.26.
97. Haber, p.109.
98. Brophy, p.9.
99. Ede, A., 'The natural defense of a scientific people: the public debate over chemical warfare in post-WWI America', *Bull. Hist. Chem.*, 2002, 27, p.128.
100. Carleton, P.W., 'Anti-dimming compositions for use in the gas mask', *Ind. Eng. Chem.*, 1919, 11, p.1105.

7. Dye of Die

1. Harries, cover.
2. Anon., 'Medical arrangements of the British Expeditionary Force', *Br. Med. J.*, 1914, 2, p.804.
3. Schick, p.145.
4. Hebden, J.C., 'Dyeing of khaki in the United States: Historical and theoretical. Address delivered before the New York Section, Society of Chemical Industry', 24 May, 1918, *Ind. Eng. Chem.*, 1918, 10, p.640.
5. Jünger, p.9.
6. Dronsfield, p.85.
7. Royal Society of Chemistry release, 20 February 2012: RSC honours *Titanic* chemist who gave up his life for female passenger.
8. Jones, G., 'Progress of the American coal-tar chemical industry during 1919', *Ind. Eng. Chem.*, 1920, 12, p.959.
9. Crossley, p.199.
10. Race, E., Rowe, F.M. and Speakman, J.B., 'The dyeing of cotton with mineral khaki - Part I. The constitution and properties of the pigments product in chromate development', *The Journal of the Society of Dyers and Colourists*, 1941, 57, p.213.

8. Caring for the Wounded

1. Anon., 'From battlefield to base hospital', *The Times*, 21 October 1914, p.7.
2. Anon., 'The way home of the wounded man – by one who has travelled it', *Br. Med. J.*, 1914, 2, p.850.
3. Anon., 'The recent operations about Ypres', *Br. Med. J.*, 1917, 2, p.194.
4. Grey Turner, G., 'The importance of general principles of military surgery', *Br. Med. J.*, 1916, 1, p.401.
5. Anon., 'The wounded from the Somme', *Br. Med. J.*, 1916, 2, p.88.
6. Lockwood, A.L., Kennedy, C.M., Bute MacFie, B., and Charles, S.F.A., 'The treatment of gunshot wounds of the abdomen', *Br. Med. J.*, 1917, 1, p.317.
7. Lett, H. and Morris-Jones, J. H., 'Surgical experiences at the Anglo-American Hospital, Wimereux, France', *Br. Med. J.*, 1915, 1, p.285.
8. Anon., 'Some impressions of a civilian at the Western Front', *Br. Med. J.*, 1916, 2, p.467.
9. Chitty, H., 'A hospital ship in the Mediterranean', *Br. Med. J.*, 1915, 2, p.529.
10. Anon., 'Casualties in the medical services', *Br. Med. J.*, 1915, 2, p.837.
11. 'Eye-witnesses', p.8.
12. Anon., 'A naval ambulance train', *Br. Med. J.*, 1918, 2, p.96.
13. Anon., 'Home hospitals and the war', *Br. Med. J.*, 1915, 1, p.439.

14. Ward, G.R., 'A dressing station in France', *Br. Med. J.*, 1915, 1, p.880.
15. Anon., 'Midwifery in the fire zone', *Br. Med. J.*, 1914, 2, p.689.
16. Anon., 'The proportions of recoveries among the wounded', *Br. Med. J.*, 1915, 2, p.414.
17. Anon., 'Surgery at the Siege of Kut', *Br. Med. J.*, 1917, 1, p.237.
18. Molyneux, E.S., 'Some observation on abdominal war surgery', *Br. Med. J.*, 1919, 1, p.737.

9. Fighting Infection

1. Remarque, p.199.
2. Pinker, p.195.
3. Bridger, p.198 (Table 23).
4. Corrigan, p.53.
5. Hoare, p.93.
6. Anon., 'Sanitation of Camps: An American View', *Br. Med. J.*, 1914, 2, p.731.
7. Osler W., 'Bacilli and bullets', *Br. Med. J.*, 1914, 2, p.569.
8. C. Brehmer Heald, 'The medical care of lines of communication at home', *Br. Med. J.*, 1914, 2, p.753.
9. G. Grey Turner, 'The importance of general principles of military surgery', *Br. Med. J.*, 1916, 1, p.401.
10. Brittain, p.195.
11. Graves, p.142.
12. Richards, p.74.
13. Anon., 'Notes: Serbia', *Br. Med. J.*, 1915, 1, p.986.
14. Woodham-Smith, p.205.
15. Anon., 'Prisoners in German camps', *Br. Med. J.*, 1916, 1, p.801.
16. Farmborough, p.160.
17. Everts, S., 'Our microbial selves', *Chem. & Eng. News.*, 13 December 2010, p.32.
18. Carey Evans, T.J., 'Clinical observations of dysentery', *Br. Med. J.*, 1917, 1, p.418.
19. Sneader, p.92.
20. Low, G.C, 'Letter: Emetine bismuth iodide in the treatment of amoebic dysentery', *Br. Med. J.*, 1918, 1, p.188.
21. Edgeworth, F.H., 'Notes on some recent cases of dysentery', *Br. Med. J.*, 1917, 1, p.362.
22. Anon., 'German experiences: treatment of typhoid fever and dysentery', *Br. Med. J.*, 1916, 1, p.464.
23. Macdonald, L., p.67.
24. Osler, W., 'The war and typhoid fever', *Br. Med. J.*, 1914, 2813, p.909.
25. Anon., 'Medical arrangements of the British Expeditionary Force', *Br. Med. J.*, 1914, 2, p.1037.

26. Anon., 'A military medical society', *Br. Med. J.*, 1915, 2, p.938.
27. Anon., 'A military medical society', *Br. Med. J.*, 1915, 2, p.905.
28. Anon., 'Typhoid fever in the German and Austrian armies', *Br. Med. J.*, 1915, 2, p.836.
29. Mehta, A., 'Aspirin', *Chem. & Eng. News.*, 20 June 2005, p.46.
30. Rogers, L., 'The variations in the pressure and composition of the blood in cholera; and their bearing on the success of hypertonic saline transfusion in its treatment', *Proc. R. Soc., Lond. B*, 1909, 81, p.291.
31. Farmborough, p.68.
32. Farmborough, p.218.
33. Anon., 'Mesopotamia dispatch', *Br. Med. J.*, 1916, 2, p.699.
34. Bowlby, A., 'British military surgery in the time of Hunter and in the Great War', *Br. Med. J.*, 1919, 1, p.205.
35. Macdonald, L., p.24.
36. Chi, C.H., Chen, K.W., Huang, J.J., Chuang, Y.C., and Wu, M.H., 'Gas composition in Clostridium septicum gas gangrene', *J. Formos. Med. Assoc.*, 1995, 94, p.757.
37. Fraser, J., Bates, H.J., 'Further observations on the treatment of gas gangrene by the intravenous injection of hypochlorous acid', *Br. Med. J.*, 1916, 2, p.172.
38. Anon., 'The war: gas gangrene', *Br. Med. J.*, 1917, 1, p.466.
39. Farmborough, p.217.
40. MacConkey, A., 'Tetanus: Its prevention and treatment by means of antitetanic serum', *Br. Med. J.*, 1914, 2, p.609.
41. Anon., 'German experiences of tetanus', *Br. Med. J.*, 1915, 2, p.906.
42. Anon., 'Tetanus in home military hospitals', *Br. Med. J.*, 1918, 2, p.415.
43. Raw, N., 'Trench nephritis: a record of five cases', *Br. Med. J.*, 1915, 2, p.468.
44. Maher, J. F., 'Trench nephritis: a retrospective perception', *American Journal of Kidney Diseases*, 1986, 7, p.355.
45. Atenstaedt, R.L., 'The medical response to trench nephritis in World War One', *Kidney International*, 2006, 70, p.635.
46. Anon, 'The war: war nephritis', *Br. Med. J.*, 1916, 1, p.251.
47. Clarke, J.M., 'Trench nephritis, its later stages and treatment', *Br. Med. J.*, 1917, 2, p.239.
48. Anon., 'Notes: Serbia', *Br. Med. J.*, 1915, 1, p.781.
49. Leneman, L., 'Medical women at war, 1914–1918, *Medical History*, 1994, 38, p.160.
50. Anon., 'The typhus epidemic at Gardelegen', *Br. Med. J.*, 1916, 2, p.623.
51. Hartcup, p.173.
52. Harrison, p.21.
53. White, p.26.
54. Hunter, W., 'The Serbian epidemics of typhus and relapsing fever in 1915:

their origin, course, and preventive measures employed for their arrest', *Proc. R. Soc. Med.,* 1920, 13 (Sect Epidemiol State Med), p.29.

55. Whitchurch Howell, B., 'Typhus in Serbia', *Br. Med. J.*, 1915, 2, p.813.
56. Patterson, K.D., 'Typhus and its control in Russia 1870–1940', *Medical History*, 1993, 37, p.361.
57. Herringham, W.P., 'Trench fever and its allies', *Br. Med. J.*, 1917, 1, p.833.
58. Atenstaedt, R.L., 'Trench fever: the British medical response in the Great War', *J. R. Soc. Med.*, 2006, 99, p.564.
59. Hughes, B., 'The causes and prevention of trench foot', *Br. Med. J.*, 1916, 1, p.712.
60. Anon., 'The war: Treatment of wounds in war. II. Trench foot', *Br. Med. J.*, 1918, 1, p.516.
61. Harrison, M., 'Medicine and the culture of command: the case of malaria control in the British Army during the two World Wars', *Medical History*, 1996, 40, p.437.
62. Balfour, A., 'Memorandum on some medical diseases in the Mediterranean war area, with some sanitary notes', His Majesty's Stationary Office, London, 1916.
63. Treadgold, C.H., 'The prophylactic use of quinine in malaria: with special reference to experiences in Macedonia', *Br. Med. J.*, 1918, 1, p.525.
64. Richards, p.74.
65. Graves, p.125.
66. Graves, p.188.
67. Graves, p.246.
68. Anon., 'The prophylaxis of venereal diseases', *Br. Med. J.*, 1917, 1, p.243.
69. McEwen, p.98.
70. Bridger, p.177.
71. Barrett, J.W., 'The management of venereal diseases in Egypt during the war', *Br. Med. J.*, 1919, 1, p.125.
72. Anon., 'Venereal disease in the Austrian Army', *Br. Med. J.*, 1917, 2, p.300.
73. Harrison, L.W., White, C.F. and Mills, C.H., 'The intramuscular or subcutaneous injection of neo-salvarsan', *Br. Med. J.*, 1917, 1, p.569.
74. Brittain, p.49.
75. Fenwick, P.C., Sweet, G.B. and Lowe, E.C, 'Two fatal cases of icterus gravis following injections of novarsenobillon', *Br. Med. J.*, 1918, 1, 448.
76. Hayes, J.G., The treatment of acute gonorrhea in men', *Br. Med. J.*, 1914, 2, p.469.
77. Fogarty, J.P., 'Treatment of acute gonorrhoea: massage-pack method', *Br. Med. J.*, 1919, 1, 245.
78. Anon., 'Means for the control of influenza', *Br. Med. J.*, 1919, 1, p.248.
79. Arthur, p.237.
80. Arthur, p.258.

81. Barry, p.4.
82. Anon., 'The influenza epidemic in the British armies in France, 1918', *Br. Med. J.*, 1918, 2, p.505.
83. Armitage, F.L., 'Note on "influenza" and pneumonia: from a field hospital', *Br. Med. J.*, 1919, 1, p.272.
84. Cooper Cole, C.E., 'Influenza epidemic at Bramshott in September-October, 1918', *Br. Med. J.*, 1918, 2, p.566.
85. Martin, N.A., 'Sir Alfred Keogh and Sir Harold Gillies: their contribution to reconstructive surgery', *J. R. Army Med. Corps*, 2006, 152, p.136.
86. Anon., 'The German Medical Congress in Warsaw', *Br. Med. J.*, 1916, 2, p.27.
87. Wansey Bayly, H., Magian, A.C. and Marshall, C.F., 'Prevention of venereal disease', *Br. Med. J.*, 1919, 2, p.863.
88. Herringham, W.P., 'Bacteriology at the front', *Br. Med. J.*, 1917, 1, p.832.
89. Anon., 'Austro-German experiences', *Br. Med. J.*, 1916, 2, p.191.
90. Walker, N.M., 'Edward Almroth Wright', *J. R. Army Med. Corps*, 2007, 153, p.16.
91. Hartcup, p.171.
92. Atenstaedt, R.L., 'The development of bacteriology, sanitation science and allied research in the British Army 1850–1918: equipping the RAMC for war', *J. R. Army Med. Corps*, 2010, 156, p.154.
93. Anon., 'The war: war, disease, and a museum', *Br. Med. J.*, 1916, 1, p.214.
94. Anon., 'Military hygiene and the efficiency of the soldier', *Br. Med. J.*, 1915, 1, p.553.
95. Macdonald, L., p.67.
96. Sassoon, p.34.
97. Anon., 'Personal hygiene', *Br. Med. J.*, 1916, 1, p.630.
98. Arthur, p.122.
99. Shipley, A.E., 'Insects and war: lice', *Br. Med. J.*, 1914, 2, p.497.
100. Parlane Kinloch, J., 'An investigation of the best methods of destroying lice and other body vermin', *Br. Med. J.*, 1916, 1, p.789.
101. Anon., 'Surgical experiences of the present war', *Br. Med. J.*, 1914, 2, p.891.
102. Lister, J., 'On the antiseptic principles in the practice of surgery', *The Lancet*, 1867, 90, p.353.
103. Lister, J., 'On the use of carbolic acid', *The Lancet*, 1867, 90, p.444, 502, 595.
104. Morison, R., 'The treatment of infected, especially war, wounds', *Br. Med. J.*, 1917, 2, p.503.
105. Morison, A.E., 'The treatment of infected war wounds by magnesium sulphate', *Br. Med. J.*, 1918, 1, p.342.
106. Smith, J.L., Drennan, A.M., Rettie, T. and Campbell, W., 'Antiseptic action of hypochlorous acid and its application to wound treatment', *Br. Med. J.*, 1915, 2, p.129.
107. Dakin, H.D., 'On the use of certain antiseptic substances in the treatment of infected wounds', *Br. Med. J.*, 1915, 2, p.318.

108. Anon., 'The war: Carrel-Dakin treatment of wounds', *Br. Med. J.*, 1917, 2, p.597.
109. Dakin, H.D., Cohen, J.B. and Kenyon, J., 'Studies in antiseptics (II): On chloramine: its preparation, properties, and use', *Br. Med. J.*, 1916, 1, p.160.
110. Dunham, E.K. and Dakin, H.D., 'Observations on chloramines as nasal antiseptics', *Br. Med. J.*, 1917, 1, p.865.
111. Sweet, J., 'Dakin's "dichloramine-T" in the treatment of the wounds of war', *Br. Med. J.*, 1917, 1, p.249.
112. Wright, A.E., 'Wound infections', *Br Med. J.*, 1915, 1, p.625.
113. Wright, 'Wound infections and their treatment', *Br. Med. J.*, 1915, 2, p.670.
114. Bowlby, A.A., Rowland, S., 'A report on gas gangrene', *Br. Med. J.*, 1914, 2, p.913.
115. Smith, S.M., 'Treatment of the wounded in the aid posts and field ambulances', *Br. Med. J.*, 1918, 2, p.127.

10. Killing the Pain

1. Anderson, H.G., 'The medical aspects of aeroplane accidents', *Br. Med. J.*, 1918, 1, p.73.
2. Brittain, p.188.
3. McEwen, p.53.
4. Roberts, J.E.H. and Statham, R.S.S., 'On the salt pack treatment of infected gunshot wounds', *Br. Med. J.*, 1916, 2, p.282.
5. Anon., 'The war: the Royal Naval Medical Service in Action', *Br. Med. J.*, 1916, 2, p.597.
6. Blakemore, P.R. and White, J.D., 'Morphine, the Proteus of organic molecules', *Chem. Commun.*, 2002, p.1159.
7. Freemantle, M., 'Morphine', *Chem. & Eng. News*, 20 June 2005, p.90.
8. Gates, M. and Tschudi, G., 'The synthesis of morphine', *J. Am. Chem. Soc.*, 1952, 74, p.1109.
9. Gortler, L., 'Merck in America: the first 70 years from fine chemicals to pharmaceutical giant', *Bull. Hist. Chem.*, 2000, 25, p.1.
10. Anon., 'The supply of intoxicant drugs to members of H.M. Forces', *Br. Med. J.*, 1917, 2, p.797.
11. Charles, R. and Sladden, A.F., 'Resuscitation work in a casualty clearing station', *Br. Med. J.*, 1919, 1, p.402.
12. Lockwood, A.L., Kennedy, C.M., Bute MacFie, B., and Charles, S.F.A., 'The treatment of gunshot wounds of the abdomen', *Br. Med. J.*, 1917, 1, p.317.
13. Sassoon, p.145.
14. Black, J.E., Glenny, E.T. and McNee, J.W., 'Observations on 685 cases of poisoning by noxious gases used by the enemy', *Br. Med. J.*, 1915, 2, p.165.

15. Anon., 'Treatment of acute dysentery', *Br. Med. J.*, 1916, 1, p.142.
16. Macdonald, A.T.I., 'Medical and surgical work as a prisoner of war', *Br. Med. J.*, 1919, 1, p.367.
17. Arthur, p.237.
18. Connor, H., 'The use of chloroform by British Army surgeons during the Crimean War', *Medical History*, 1998, 42, p.161.
19. Hoare, p.123.
20. Anon., 'A motor operating theatre', *Br. Med. J.*, 1915, 1, p.129.
21. Reisch, M., 'Ether', *Chem. & Eng. News*, 20 June 2005, p.62.
22. Silk, J.F.W., 'Anaesthetics: a modification of the open ether method', *Br. Med. J.*, 1919, 1, p.635.
23. McCardie, W.J., 'A method of anaesthetizing soldiers', *Br. Med. J.*, 1917, 1, p.508.
24. Mills, A., 'The administration of anaesthetics to soldiers', *Br. Med. J.*, 1918, 2, p.343.
25. Lawson, J.I.M., 'Ethyl chloride', *Brit. J. Anaesth.*, 1965, 37, p.667.
26. Good J.P., 'Anaesthetics in a British hospital with the Serbian Army', *Brit. Med. J.*, 1919, 1, p.667.
27. Freemantle, M., 'A candle burning brightly', *Chem. & Eng. News*, 11 September 2000, p.34.
28. Davy, H., 'Researches chemical and philosophical, chiefly concerning nitrous oxide, or dephlogisticated nitrous air, and its respiration', J. Johnson, London, 1800, p.465.
29. *Ibid.*, p.556.
30. Lett, H. and Morris-Jones, J. H., 'Surgical experiences at the Anglo-American Hospital, Wimereux, France', *Brit. Med. J.*, 1915, 1, p.285.
31. Ruetsch, Y.A., Boni, T. and Borgeat, A., 'From cocaine to ropivacaine: the history of local anesthetic drugs', *Curr. Top. Med. Chem.*, 2001, 1, p.175.
32. Rayner-Canham, M.F. and Rayner-Canham, G.W., 'British women chemists and the First World War', *Bull. Hist. Chem.*, 1999, 23, p.20.
33. Hartcup, p.168.
34. Creese, M.R.S., 'Martha Annie Whiteley (1866–1956): chemist and editor', *Bull. Hist. Chem.*, 1997, 20, p.42.

11. A Double-Edged Sword

1. Carrel, A., 'Science has perfected the art of killing – why not of saving?' *Surgery, Gynecology, & Obstetrics*, 1915, 20, p.710.
2. Dakin, H.D., 'Biochemistry and war problems', *Brit. Med. J.*, 1917, 1, p.833.
3. Freemantle, M., p.27.

Bibliography

Abbott, A., *Airships*, Shire Publications, Princes Risborough, 1991.

Adams, M.R. and Moss, M.O., *Food Microbiology*, 2nd Edn, The Royal Society of Chemistry, Cambridge, 2000.

Ainslie, G.M., *Hand Grenades: A Handbook on Rifle and Hand Grenades* (reprint of 1917 original edition), Naval & Military Press, Uckfield, 2008.

Akhavan, J., *Chemistry of Explosives*, RSC Publishing, Cambridge, 2004.

Arthur, M., *Last Post*, Weidenfeld & Nicolson, London, 2005.

Auld, S.J.M., *Gas and Flame in Modern Warfare*, George H. Doran Company, New York, 1918.

Badger, G., *Collecting photography*, Mitchell Beazley, London, 2003.

Barry, J.M., *The Great Influenza: The Story of the Deadliest Pandemic in History*, Penguin, London, 2004.

Barton, P., Doyle, P. and Vandewalle, J., *Beneath Flanders Fields: The Tunnellers' War 1914–1918*, McGill-Queen's University Press, 2005.

Botting, D., *The U-boats*, Time Life Books, Amsterdam, 1979.

Bridger, G., *The Great War Handbook*, Pen & Sword, Barnsley, 2009.

Brittain, V., *Testament of Youth*, Fontana Paperbacks, Virago, London, 1979.

Brophy, L.P. and Fisher G.J.B., *The Chemical Warfare Service: Organizing for War*, Center of Military History, Washington DC, 2004.

Brown, G.I., *Explosives: History with a Bang*, The History Press, Stroud, 2010.

Charles, D., *Between Genius and Genocide: The Tragedy of Fritz Haber, Father of Chemical Warfare*, Pimlico, London, 2006.

Clarke, D., *British Artillery 1914–19: Field Army Artillery*, Osprey, Oxford, 2004.

Coffey, P., *Cathedrals of Science: The Personalities and Rivalries That Made Modern Chemistry*, Oxford University Press, Oxford, 2008.

Corrigan, G., *Mud, Blood and Poppycock*, Cassell, London, 2003.

Croddy, E., with Perez-Armendariz, C. and Hart, J., *Chemical and Biological Warfare: A Comprehensive Survey for the Concerned Citizen*, Copernicus Books, New York, 2002.

Crossley, R.S., *Accrington Captains of Industry*, Wardleworth, Accrington, 1930.

Crowther B. and Freemantle, M., *Experiments and Investigations in Chemistry*,

Oxford University Press, Oxford, 1989.
Davis, T.L., *The Chemistry of Powder and Explosives*, Angriff Press, Hollywood, Calif., 1972.
Dronsfield, A. and Edmonds, J., *The Transition from Natural to Synthetic Dyes*, John Edmonds, Little Chalfont, 2001.
Eaton, J.P. and Haas, C.A., *Titanic: Triumph and Tragedy*, Patrick Stephens, Wellingborough, 1986.
Emsley, J., *The Shocking History of Phosphorus: A Biography of the Devil's Element*, Macmillan, London, 2000.
Evans, R., *Gassed*, House of Stratus, London, 2000.
Eye-witnesses, *The War on Hospital Ships*, T. Fisher Unwin, London, 1917.
Farmborough, F., *Nurse at the Russian Front:* A Diary 1914-1918, Constable, London, 1974.
Faulks, S., *Birdsong*, Vintage, London, 1994.
Forty, G., *Royal Tank Regiment*, Guild Publishing, London, 1989.
Freemantle, M., *Chemistry in Action*, 2nd Edn, Macmillan Press, 1995.
Freemantle, M.H. and Tidy, J.G., *Essential Science: Chemistry*, Oxford University Press, Oxford, 1983.
Graves, R., *Goodbye to All That*, Penguin Books, London, 1960.
Haber, L.F., *The Poisonous Cloud, Chemical Warfare in the First World War*, Clarendon Press, Oxford, 1986.
Hall, M., *In Enemy Hands: A British Territorial Soldier in Germany 1915–1919*, Tempus, Stroud, 2002.
Hamilton, D.T., *High-Explosive Shell Manufacture*, The Industrial Press, New York, 1916.
Hanson, N., *First Blitz*, Corgi, London, 2009.
Harries, M. and S., *The War Artists*, Michael Joseph, London, 1983.
Harris, R. and Paxman, J., *A Higher Form of Killing: The Secret History of Chemical and Biological Warfare*, Arrow, London, 2002.
Harrison, M., *War and Medicine*, Larner, M., Peto, J., and Monem, N. (eds), Black Dog Publishing, London, 2008.
Hartcup, G., *The War of Invention: Scientific Developments, 1914–18*, Brassey's Defence Publishers, 1988.
Haythornthwaite, P.J., *The World War One Source Book*, BCA, London, 1992.
Hellemans, A., Bunch, B., *The Timetables of Science: A Chronology of the Most Important People and Events in the History of Science*, Simon and Schuster, New York, 1988.
Hitler, A., *Mein Kampf*, translated by Murphy, J., Hurst & Blackett, London, 1939.
Hoare, P., *Spike Island: The Memory of a Military Hospital*, Fourth Estate, London, 2002.
Hogg, I.V., and Thurston, L.F., *British Artillery Weapons & Ammunition 1914–1918*, Ian Allan, Shepperton, 1972.

Holmes, R., *The Western Front*, BBC Worldwide, 1999.

Holt, T. & V., *Pocket Battlefield Guide to Ypres & Passchendaele*, Pen & Sword Military, Barnsley, 2008.

Jones, S., *World War I Gas Warfare Tactics and Equipment*, Osprey, Oxford, 2007.

Jünger, E., *Storm of Steel*, Penguin Books, London, 2004.

Lee, J., *The Gas Attacks Ypres 1915*, Pen & Sword, Barnsley, 2009.

Lewis, D., *The Man Who Invented Hitler*, Headline, London, 2003.

McAuliffe, C.A., *The Chemistry of Mercury*, Macmillan Press, London, 1977.

McDonald, D., and Hunt, L.B., *A History of Platinum and its Allied Metals*, Johnson Matthey, London, 1982.

Macdonald, G., *Camera: Victorian Eyewitness*, B.T. Batsford, London, 1979.

Macdonald, L., *The Roses of No Man's Land*, Papermac, London, 1984.

McEwen, Y., *'It's a Long Way to Tipperary' – British and Irish Nurses in the Great War*, Cualann Press, Dunfermline, 2006.

Macksey, K., *The Penguin Encylopedia of Weapons and Military Technology*, Penguin, London, 1995.

MacLeod, R. and Johnson J.A. (eds), *Frontline and Factory: Comparative Perspectives on the Chemical Industry at War, 1914–1924*, Springer, Dordrecht, 2006

Marshall, A., *Dictionary of Explosives*, P. Blakiston's Son & Co., Philadelphia, 1920.

Nye, M.J., *Science in the Provinces: Scientific Communities and Provincial Leadership in France, 1860–1930*, University of California Press, Berkeley, 1986.

O'Neil, M.J. (ed.), *The Merck Index*, 13th edn, Merck & Co Inc, Whitehouse Station, NJ, 2001.

Palazzo A., *Seeking Victory on the Western Front: The British Army and Chemical Warfare in World War I*, University of Nebraska Press, Lincoln and London, 2000.

Partington, A. (Ed.), *The Oxford Dictionary of Quotations*, 4th Edn, Oxford University Press, Oxford, 1996.

Pinker, S., *The Better Angels of Our Nature: The Decline of Violence in History and its Causes*, Allen Lane, London, 2011.

Prentiss, A.M., *Chemicals in War: A Treatise on Chemical Warfare*, McGraw-Hill, New York, 1937.

Remarque, E.M., *All Quiet on the Western Front*, Jonathan Cape, London, 1994.

Richards, F., *Old Soldiers Never Die*, Faber and Faber, London, 1964.

Russell, M. S., *The Chemistry of Fireworks*, RSC Publishing, Cambridge, 2000.

Ryan, T.A., Ryan, C., Seddon, E.A. and Seddon, K.R., *Phosgene and Related Carbonyl Halides*, Elsevier, Amsterdam, 1996.

Sassoon, S., *Memoirs of an Infantry Officer*, Faber and Faber, London, 1965.

Saunders, A. *Weapons of the Trench War 1914–1918*, Sutton Publishing, Stroud, 1999.

Schick, I.T., *Battledress: The Uniforms of the World's Great Armies 1700 to present*, Weidenfeld & Nicolson, London, 1978.

Sheffield G. (ed.), *War on the Western Front*, Osprey, Oxford, 2008.

Silver, J., *Chemistry of Iron*, Chapman & Hall, London, 1993.

Sneader, W., *Drug Discovery: A History*, John Wiley & Sons, 2005.

Stevenson, D., *1914–1918 The History of the First World War*, Penguin Books, London, 2005.

Storey, N. and Housego, M., *Women in the First World War*, Shire Publications, Oxford, 2010.

Taylor, A.J.P., *The First World War: An Illustrated History*, Penguin Books, Harmondsworth, 1966.

Tuorinsky, S.D., (Ed.), *Medical Aspects of Chemical Warfare*, Office of the Surgeon General at TMM Publications, Washington DC, 2008.

Travis, A.S., Schröter, H.G., Homburg, E. and Morris, P.J.T. (eds), *Determinants in the Evolution of the European Chemical Industry, 1900–1939*, Kluwer Academic Publishers, Dordrecht, The Netherlands, 1998.

Vilensky, J.A., *Dew of Death: The Story of Lewisite, America's World War I Weapon of Mass Destruction*, Indiana University Press, Bloomington, 2005.

Wakefield, A., *Christmas in the Trenches*, Sutton Publishing, 2006.

Warner, P., *Passchendaele: The Story Behind the Tragic Victory of 1917*, Sidgwick & Jackson Limited, London, 1987.

War Office, *Treatise on Ammunition*, 10th Edn, The Naval & Military Press, Uckfield, East Sussex, and The Imperial War Museum, London, 1915.

Westwell, I., *The Complete Illustrated History of World War I*, Lorenz Books, London, 2008.

White, M., *The Fruits of War: How Military Conflict Accelerates Technology*, Simon & Schuster UK, London, 2005.

Willmott, H.P., *World War I*, Dorling Kindersley, London, 2008.

Woodham-Smith, C., *Florence Nightingale*, Constable, London, 1950.

Index

ABE fermentation 73
Abel, Frederick 71
Acetone 18, 71–73, 123, 204
Acetylene 18, 136–137
Acriflavine 197
Acrolein 121–123
Adrenalin 181, 189
Adrian helmet 92
Aerial photography 110–111
Airships 100–102, 211
Alizarin 157
Alkaloids 162, 175, 183, 189, 200–202, 207–209
Alloys 12–13, 17–18, 34–35, 39–40, 86, 89, 92–96, 98–103, 107, 211, 213
Alumina 102–103
Aluminium 17–18, 35, 39–40, 46–47, 52–53, 57, 77–78, 81, 95, 98–103
  Aluminium chloride 137
  Aluminium hydroxide 102
  Aluminium oxide 102
Amatol 34, 39, 76–77
Ambulances 159–166, 177, 199
  Ambulance trains 163–165, 168, 190
  Field ambulances 160–162, 165–166, 168, 175
American Chemical Society 12, 19, 21, 83, 149
American Civil War 109
American Expeditionary Force (AEF) 61, 131–133, 136–137, 148, 188
American University Experiment Station 148–150
Ammonal 53, 55–56, 63–64, 66, 68, 76–78, 207, 213
Ammonia 16, 21–22, 25, 34, 74, 84–86, 91, 138, 141, 207
  Ammoniated mercuric chloride 193
  Ammonium bromide 189
  Ammonium chloride 138
  Ammonium nitrate 19, 34, 53, 61, 76–78, 81, 206, 213
  Ammonium Nitrate Fuel Oil (ANFO) 76
  Ammonium picrate 34, 74
  Ammonium sulfate 16
Amoebic dysentery 174–175, 183
Amyl alcohol 67
Anaesthetics 168, 203–209, 212
  General anaesthetics 203–207
  Local anaesthetics 208–209
  Spinal anaesthesia 208
Analgesics 162, 168, 176, 184, 199–203, 212
Aniline 37, 74, 155
Anthracene 157
Anti-aircraft shells 47
Antibiotics 187
Anti-flash additive 75
Antimony 40, 99–100
  Antimony trisulfide 38, 79
Antipyretics176
Antiseptics 21, 125, 159–160, 163–164, 168, 179–180, 184, 187, 191, 194–198, 203,
  212–213
Appert, Nicolas 105
Archduke Franz Ferdinand of Austria 54
Archer, Frederick Scott 108
Armistice 116, 120, 132, 137, 189
Armour 12, 17, 21, 40, 93, 213
  Armour plate 91
  Krupp cemented armour 92
  Steel armour 92
Armour-piercing shells 39, 46, 74, 95
ARS mask 143
Arsenic 100
Arsenic compounds 43, 118–119, 124, 186
  Arsenic trichloride 126, 131, 136–137
Arsenicals 124, 136–137, 148, 187, 197
Artillery shells 27–49
Aseptic procedures 160, 194
Aspirin 176, 184, 189, 203, 208–209
Asquith, Herbert 29
Atropine 189, 207, 209
Auld, Samuel J. M. 139–140
Autochrome 108
Bacillary dysentery 174–175
Bacilli 174, 176
Bacteria 73, 104, 170, 173–174, 178–182, 189–190, 194–195
  Anaerobic bacteria 178–190
Baekeland, Leo 11, 16, 23
Bakelite 11
Balfour, Arthur 73
Ballistite 37, 70–71
Barbed wire 17, 28, 91, 213
Barbiturate 180, 203
Barium compounds 48, 98
  Barium chlorate 49, 98
  Barium nitrate 48, 56, 68
  Barium platinocyanide 112
Barley mask 139
Battle of Arras 42, 44, 202
Battle of Aubers Ridge 29
Battle of Bolimów 41
Battle of Cambrai 115
Battle of Champagne 154
Battle of Courtrai 116
Battle of Hill 60 66
Battle of Jutland 45, 72, 200
Battle of Loos 44, 117, 120, 123, 126
Battle of Masurian Lakes 28
Battle of Messines 27, 78
Battle of Mons 200
Battle of Neuve Chapelle 29, 43, 126
Battle of St Quentin Canal 132–133
Battle of Passchendaele 204
Battle of the Aisne, 1914 154, 160
Battle of the Frontiers 154, 177
Battle of the Marne, 1914 160
Battle of the Somme
  (1916) 162, 164, 179, 192
  (1918) 179
Battle of Verdun 27, 42, 128, 151
Battle of Ypres
  (1915) 14, 44, 47, 115, 117, 120, 126
  (1917) 42, 47, 57, 132, 135, 161
Bauxite 23, 102
Bayer 129, 135, 177, 202
Bayer, Karl 102
Becquerel, Henri 113
Benzyl bromide 41, 122
Benzyl iodide 123
Berthollet, Claude 80
Bessemer converter 89–92
Bessemer, Henry 89
Beta-eucaine 208–209
Biocomposite 33
Bipp paste 195
Birkeland-Eyde process 85
Bismuth compounds 175, 195
Black Veil Respirator 139–140
Blast furnace 88, 90
Blastine 75
Blasting explosive 68–70, 74, 76–77
Bleach 125, 156
Bleaching powder 130, 145, 192, 196
Blimps 102
Blistering agents, see Vesicants
Blood agents 118, 123, 130–131
Bloomery 88
Blue Cross shells 43, 119, 124
Blue Star gas 128
Bn-Stoff 41, 122
Boer Wars 74, 109, 153, 170, 189
Booster explosives 30–31, 34–35, 39, 60, 63, 74, 80, 81

Boric acid 184, 195–196
Bosch, Carl 16
Bowlby, Sir Anthony 178–179, 197
Brass 25, 32, 35, 37–38, 40, 53, 55, 86, 95–98
Breech-loading weapons 32, 37–39, 72, 152
Bridging explosive, see Booster explosives
Brilliant green 197
Brimstone, see Sulfur
British chemical industry 14
British Chemical Warfare Committee 147
British Expeditionary Force (BEF) 19, 29, 55, 110, 147, 160, 165
British Large Box Respirator 142–143, 177–178, 197–198, 200
British Ministry of Munitions 62, 73, 85, 147
British Phenate Helmet 140–141
British Smoke Hood 139
Brittain, Vera 115, 171, 187, 199–200
Brodie helmet 92, 106
Bromine 42, 122–123
Bromine compounds 41, 80, 107, 122–123, 131
Bromoacetone, see B-Stoff
Bromomethylethyl ketone, see Bn-Stoff
Bronze 12, 34, 39, 86, 95–96, 98
Brown powder 66
B-Stoff 41, 122–123
Bullets 18, 98–100
Buckingham bullet 101
Dumdum bullet 99
Pointed rifle bullet 39, 98–100
Pomeroy bullet 101
S rifle bullet 98
Shrapnel bullet 30, 100
Bursting charge 30–34, 38–43, 46–48, 53–56, 60, 65, 69, 74, 77–79, 81, 97–98, 130
Calcium compounds 45, 72, 85, 97, 143
Calcium carbide 18, 85
Calcium carbonate 33, 72, 85, 88, 194, 196
Calcium chloride 97
Calcium cyanamide 85
Calcium hypochlorite 130, 135, 145, 192–193, 204
Calcium oxide 72, 85, 149, 166, 195
Calcium tungstate 112
Calomel 185, 189
Calotypes 107–109
Camouflage 47, 152, 154, 213
Camouflet 66, 68
Camphor 37, 47, 70–71, 74, 108, 184, 189, 202
Carbolic acid, see Phenol
Carbon 12–13, 15, 17, 24–25, 33, 36, 53, 76–77, 88–93, 97, 103, 149
Carbon dioxide 36–37, 63, 65, 68, 93, 103, 138
Carbon monoxide 36–37, 68, 75, 103, 127, 129
Carbonyl chloride, see Phosgene
Carrel, Alexis 196, 212
Cartridges 24, 31–32, 37–38, 43, 49, 56, 61, 65, 70–72, 79, 95–96, 106
Cassia oil 185
Cassiterite, see Tin dioxide
Cast iron 33, 53, 55, 88–89, 93, 100, 123
Cast iron shells 33, 38–40, 47, 123
Casualty clearing stations 93, 151, 160–164, 166, 168, 171, 179, 190, 192, 197, 200, 202–204, 212
Casualty statistics 59, 120–121, 169–170, 180–182, 189, 199
Catalysis, catalysts 16, 83–85, 91, 137
Caustic soda, see Sodium hydroxide
Celluloid 13, 47, 67, 108–109, 112
Celluloid roll film 13, 108–109
Cellulose nitrate, see Guncotton
Chalcopyrite 96
Charcoal 12, 35, 64–66, 76–77, 79, 88, 138, 142–144, 149
Activated charcoal 142–143
Chemical industry 60, 117, 156
Chemical warfare 14, 20–21, 93, 115–150
Chemical warfare agents 14, 18, 21, 30, 41, 43–44, 93, 116–137, 144, 146–150
Chemical Warfare Service 21, 117, 134, 142–143, 145, 148–150
Chemists' War 11–25, 214,
Chemotherapy 136, 187
Chile saltpetre, see Sodium nitrate
Chloral hydrate 180
Chloramines 196
Chlorine 14, 18, 21, 24–25, 42–44, 93, 115–120, 124–132, 138–145, 192, 213
Chloroacetone 57, 122–123
Chloroacetophenone 123
Chloroform 130–131, 163, 193, 199, 203–208
Chloromethyl chloroformate 41, 122
Chloropicrin 42–43, 118, 126, 129–132, 134, 142
Chlorosulfonic acid 45
Chlorvinyldichlorarsine 136–137
Choking agents 117–119, 124, 128, 214
Cholera 177, 190
Chrome yellow 155
Chromium 17, 92, 94
Chromium compounds 155–158
Cine camera 109
Cloud gas 14, 44, 118–120, 124, 126–127, 129–131, 169, 211
CN gas, see Chloroacetophenone
Coal 23–24, 45, 75, 84, 88–89
Coal tar 14–15, 23–24, 46, 74–75, 77, 88, 140, 155–157, 191–194, 211, 213–214
Coal-boxes 28
Cocaine 208
Cochineal 155
Codeine 162, 189, 201–202
Collodion 70–72, 74, 108
Collodion process 108
Colour photography 107–109
Combustion 36–37, 45, 63, 65–66, 76
Common lyddite shells 39
Common shells 34, 38, 65
Common-pointed shells 39
Composition exploding, see Tetryl
Composition priming 66
Contact process 83–86
Copper 12, 18, 20, 23, 34–35, 39, 86, 94–99, 102, 107:
Copper carbonate 49, 86, 98
Copper chloride 86
Copper iron sulfide 96
Copper-nickel alloy 99, 102
Cordite 31, 36–38, 56, 60, 62–63, 65, 69, 71–73, 75, 200
Cordite factories 60, 62, 71–73
Cordite MD 72
Cordite RDB 72
Cordus, Valerius 205
Corrosion 46, 86, 92, 95, 105
Corrugated iron 91, 93, 97, 98, 213
Creosote 189, 193
Cresols 167, 192–194
Crimean War 20, 109–110, 117, 170, 172, 189, 204
Crucible steel 89–91
Cryolite 103
Crystal violet 197
C-Stoff, see Chloromethyl chloroformate
Curie, Marie 112–113
Curie, Pierre 113
Cutch 154
Cyanogen bromide 131
Cyanogen chloride 131
Daguerre, Louis 20, 107
Dakin, Henry Drysdale 195–196, 212
Dardanelles 171, 186
Darzens, Georges 74
Daughters of the American Revolution 86–87
Davy, Humphry 125, 127
Davy, John 127
Death tolls, see Casualty statistics
Debridement 179–180
Deflagration 63
Detonation 30–31, 34–35, 40, 52, 54, 63, 66, 74, 77, 95
Detonators 34–35, 43, 53, 55–57, 69, 75, 81, 106
Dew of death 136–137, 149
Dewar, James 71
Diarrhoea 171–172, 174–177, 203
Diamorphine, see Heroin
Dianisidine chlorosulfate 42, 126
Diatomaceous earth 69–70, 142, 149
Diatomite, see Diatomaceous earth
Dichloramine–T 196
Diethyl ether 162, 205–206
Digitalis 183, 189
Dinitronaphthalene 78
Dinitrotoluene 75
Diphenylamine 67
Diphenylchlorarsine 43, 118–119, 124
Diphosgene 42, 119, 126, 128–130, 132
Disinfectants 20, 166–168, 170, 191–194, 212
Dope 18, 72, 102
Double-based propellant 70–71

Dover's powder 164, 189, 203
Dressing stations 159, 161–162, 165–167, 199, 203–204
D-Stoff 127
Dunnite 74
Duralumin 101–102
Dyes 151–158
Coal tar dyes 15, 23–24, 213
Fast dyes 154
Synthetic dyes 13–14, 23–25, 88, 125, 127, 154, 213
Vat dyes 156
Dynamite 24, 61, 69–70, 101
Dysentery 172, 174–176, 183–184, 203
Eastman, George 13, 108
Ehrlich, Paul 187, 197
Electric arc furnace 17–18, 90–91, 103
Electrochemistry 16–18, 213
Electrolysis 17–18, 103, 113, 125
Electroplating 105–106
Emetine 175, 183
Enamel 92
Enteric fevers 175–176
Enzymes 138
Epsom salts 175, 189
Ethanol 67, 70, 72–73, 79, 96, 108, 122, 135, 163, 179–180, 198, 205–206
Ether 205
Ethyl alcohol, see Ethanol
Ethyl bromoacetate 57, 122–123
Ethyl chloride 203, 206
Ethyl ether 67
Ethyl iodoacetate 42, 123
Eupad 195
Eusol 195, 198
Exploders, see Booster explosives
Explosive D 74
Falkenhayn, General Erich von 27
Faraday, Michael 117
Farmborough, Florence 173, 177, 179
Fermentation 73
Ferrous metals 89, 94
Fertilisers 15–16, 22, 25, 76, 84, 206, 213
Field artillery 31, 41
Filling factories 30, 60
Film 13, 22, 108–112
First field dressings 160–161
Flame test 97
Flamethrowers 46
Food preservation 103–105
Formaldehyde 141
Fox Talbot, William Henry 107
Fragmentation bursting charge 15, 30, 34, 38
Franck, James 126
French chemical warfare programme 146
French, Sir John 29
Fungi 104, 173–174, 187
Fuses 30, 34–36, 65, 79, 81, 95–96, 98, 100, 106, 211
Bickford safety fuse 55
Graze fuse 35
Igniferous time fuse 35
Impact fuse, see Percussion fuse
Mechanical time fuse 35
Percussion fuse 34–40, 55, 95
Quickmatch fuse 67
Slow match fuse 54
Time fuse 34–35, 40, 48, 55
Galena 96
Gallipoli 160, 163–164, 175, 178
Galvanization 93
Gangrene 177–179, 184
Garancin 157
Gas cylinders 44, 91
Gas gangrene 177–179, 197
Gas shells 32–34, 41–44, 46, 60, 115–116, 120, 131–133, 135–136, 148–149, 211
Gas delivery systems 43–44
Gatty, Frederick 156–158
Gatty's mineral khaki dye 156–158
Gelatine dry plate 108
Gelignite 70, 108
Gerhardt, Charles 176
German chemical industry 14–16
Germicides 194–195
Glauber, Johann 76
Godlee, Sir Rickman 194
Gonorrhoea 186–187
Green Cross shells 42, 119, 124
Green Star gas 126
Grenades 51–57, 67, 69, 77, 79, 122–123, 147, 211
Gretna 60, 62, 71
Grignard, Victor 128, 133, 146
Guanidine 80
Guncotton 35–37, 56, 59–61, 64–65, 67–73, 75, 81, 83, 108
Gunmetal 34, 96–98
Gunpowder 12, 30–31, 34–40, 46–49, 54–55, 60–61, 63–68, 78–79, 97, 152, 212
Guns 17, 19–21, 27–29, 31–33, 37, 39, 60, 65–66, 68, 70, 72, 79, 91, 93–94, 96, 101, 120, 211
Gunshot wounds 93, 162, 168, 200, 202, 204
H1N1 virus 188
Haber, Fritz 16, 22, 122, 126, 135, 146
Haber, Ludwig 122, 145–146, 148
Haber-Bosch process 16, 21, 25, 84–85, 91
Hadfield's steel 92–93
Hague Conventions 99, 117, 121, 123
Hahn, Otto 126
Hales rifle percussion grenade 56
Hall, Charles Martin 103
Hall-Héroult cell 103
Halogens 42, 107, 122, 125, 128, 175
Harrison's Tower, see British Large Box Respirator
Hartley, Harold 84–85
Helium 101, 121
Hematite 88
Hennell, Henry 96
Heroin 189, 201–202
Héroult, Paul 90, 103
Herschel, Sir John 107–108
Hertz, Gustav 126
Hexamine 141–142, 144, 187, 189
High explosives 15, 19, 29–30, 34, 36–41, 43, 46, 53, 55–56, 60, 62–63, 69, 74–78, 80,
121, 130, 136
Hindenburg Line 133
Hitler, Adolf 116
Höchst 129
Hooge 47, 49, 68
Horse chestnuts 73
Hospitals 110, 170, 172, 181–183, 203–204, 207
Base hospitals 160–164, 178–179
Home hospitals 160, 165, 167
Hospital ships 159–160, 163–164, 168
Hospital trains, see Ambulance trains
Howard, Edward 79
Howitzers 28, 31–33, 37, 39, 41, 43, 46, 70, 79, 92, 97, 214
HS gas 135
Hydrochloric acid 45–46, 125, 139, 206
Hydrogen 16, 18, 47–48, 84, 91, 101, 125, 179, 211
Hydrogen cyanide 123, 131, 141
Hydrogen peroxide 164, 168, 179, 194, 197
Hydrogen sulfide 65, 126
Hygiene 14, 171, 173, 184, 190–193, 212, 214
Hyoscine 162, 207
Hypertonic saline 177, 197
Hypo helmet 139–140
Hypochlorite of lime, see Calcium hypochlorite
Hypochlorous acid 125, 139, 179, 195, 213
Illumination shells, see Star shells
Incendiary bombs 101, 106
Incendiary shells 46–48, 67, 81
Indigo 156
Industrial revolution 87–88, 105
Inert gases 121–122
Influenza, see Spanish flu
Inorganic compounds 76, 79, 112, 194, 206, 211
Iodine 20, 107, 123, 163, 179, 194–195
Iodoacetone 123
Iodoform 193, 195
Iron, see Cast iron, Pig iron, Wrought iron
Iron oxide 47, 57, 81, 86, 94, 101–102
Iron picrate 39
Iron pyrite 83, 94
Jam tin bombs 55, 106
Jellicoe, Admiral Sir John 45
Jünger, Ernst 28, 47, 65, 154
Kaiser Wilhelm Institute 135, 146
Keogh, Sir Alfred 170, 191
Khaki 151–158, 213
Kieselguhr, see Diatomaceous earth
Krupp company 28, 35, 91–92, 95
K-shells 41
K-Stoff, see Chloromethyl chloroformate
Kunckel, Johann 79
Lachrymatory agents 41–42, 56–57, 115, 118–119, 121–123, 126, 130–132, 134, 139,
141–144
Langemarck 44, 126
Laudanum 202–203
Lavoiser, Antoine Laurent 12
Lead 35, 40–41, 44, 46, 95–96, 98–100, 102, 107
Lead–antimony alloy 100
Lead azide 81
Lead chromate 155
Lead sulfate 46
Lead sulfide 96
Lead chamber process 84
Lévy, René 155

Lewis, Winford Lee 137
Lewisite 137, 149
Lice 171, 174, 181–183, 193
Lime, see Calcium oxide
Limestone 85, 88, 90
Lines of communication 160, 165, 186, 190
Lister, Joseph 190
Livens projector 43–44
Lloyd George, David 19, 28–29
Lochnagar Crater 78
Louse-borne diseases 181–183
Low explosives 30–31, 34, 37, 63–64, 67
Lumière brothers 108–109
Lyddite 38–40, 55, 60, 63–64, 73–75, 130, 195
Lysol 167, 194
M2 fabric mask (French) 141, 143
Maddox, Richard Leach 108
Magnesium 18, 47–49, 81, 98, 101, 128
  Magnesium silicate 193
  Magnesium sulfate 84, 175, 180, 194–195
Malachite green 195
Malaria 183–185
Manganese 33, 92–96, 101
  Manganese bronze 95–96
  Manganese dioxide 96, 125
Max Planck Society 146
Maxwell, James Clerk 108
Mealed powder 35, 40, 49, 65, 67–68
Melinite 15, 74
Merck 201
Mercury 20, 35, 79, 96–97, 107, 113
  Mercury fulminate 35, 38, 43, 55, 79, 96
  Mercury salts 97, 185, 187, 192
Metalloids 92, 100
Metallurgy 17–18, 22
Microbes, see Microorganisms
Microorganisms 104–105, 164, 170–171, 173, 182–184, 187, 191–192, 194–195, 197
Military uniforms 151–158
Mills bombs 51–54, 67, 212
Mills, Sir William 52
Molybdenum 17, 93
Mordants 155
Morphine 160–162, 184, 189, 199–203, 207
Mortars 31–32, 37, 41, 43–44, 47, 77, 81, 123, 128, 130, 147, 212, 214
Mosquito repellent 185
Mosquitoes 185
Moulton, Lord John Fletcher 77
Munitionettes 62
Munitions factories 19–20, 29, 52, 61–62
Muriatic acid 125, 206
Mustard gas 42, 93, 115–119, 131–136, 144–145, 213
Muzzle-loading weapons 31–32, 37, 43, 152
Naphthalene 46, 78, 193
Napoleon I, Emperor of France 104, 176
Naval blockade 16, 23–25, 68, 73, 84, 95
Neosalvarsan 187
Nephritis, see Trench nephritis
Nickel 17, 23, 33, 92–96, 98–99, 102
  Nickel compounds 93, 141
Niépce, Nicéphore 107
Nieuwland, Julius Arthur 136–137
Nightingale, Florence 172, 189
Nitramines 80
Nitration 67–68, 75, 80
Nitric acid 15–16, 25, 36, 56, 68, 74, 76, 79–80, 83–86, 107, 211
Nitroanilines 155–156
Nitrocellulose, see Guncotton
Nitrogen fixation 16
Nitrogen mustard 135–136
Nitroglycerine 37, 53, 59, 69–73, 75–76, 83, 181, 213
Nitroguanidine 80–81
Nitrous oxide 206–207
Nobel, Alfred 69
Noble metals 86
Non-ferrous metals 92, 94–97
Novocaine 208–209
Obscurant smokes 45–46, 57
Obturation 38
Oil of bergamot 185
Omnopon 162–163, 202
Open hearth process, see Siemens–Martin process
Ophorite 47, 81
Opium 162, 189, 201–203
Ores 12, 25, 88, 93–94, 96, 100, 102
Organic compounds 13, 76, 155, 211
Ostwald process 16, 25, 85–86
Ostwald, Friedrich Wilhelm 16, 85
Owen, Wilfred 124
Oxidation 36, 65, 68, 75, 86
Oxygen 12, 16, 36, 45, 68, 75–76, 79, 83–85, 88, 93, 102–103, 118, 127, 131, 155–156, 162, 178, 205–208
P2 mask (French) 141
P mixture 38, 65
Pad respirator 139
Papaverine 202
Paraffin 47, 81, 150, 193, 195
Paratyphoid 175–176, 191
Parkes, Alexandre 13
Pasteur, Louis 105
Pathogenic microorganisms 170, 173–174, 176, 182, 184, 191, 194–195, 197
Pebble powder 38, 65–66
Pentlandite 93
Percussion cap 31, 38, 49, 53–54, 79
Perkin, William Henry 13, 15, 122–123, 155
Persistency of poisonous gases 125, 129–131, 133–134, 145–147
Petite Curies 113
Pewter 107
PH helmet 141
Pharmaceuticals 13, 15, 21–25, 28, 88, 125, 168, 170, 201–202, 212, 214
P-helmet, see British Phenate Helmet
Phenol 15, 23–24, 74, 140–141, 155, 164, 191–195
Phillips, Peregrine 83
Phosgene 18, 42–43, 57, 93, 116–120, 126–134, 140–142, 145, 187, 205, 213
Phosphate rock 23, 45–46
Phosphor bronze 35, 55, 95, 98
Phosphorus 18, 25, 44–47, 90, 95, 101–102, 106, 117, 211
  Red phosphorus 45
  White phosphorus 45–47, 57, 101
Photography 13–14, 20–23, 28, 86, 88, 107–113, 139
Picric acid 15, 19, 24, 34, 39, 41, 55, 59, 73–75, 77, 130, 162–163, 184, 195, 198
Picric powder 34, 60, 74
Pig iron 88–90
Pigments 20, 25, 131, 152, 155, 157–158
Pioneer Regiment 126
Plastics 11, 13, 25, 67
Platinotypes 86
Platinum 16, 83–87, 97, 106
Playfair, Lyon 117
Pneumonia 188
Poison gas, see Chemical warfare agents
Porter, Lord George 214
Porton Down 147–148
Potassium compounds 80, 97, 107–108, 143–144, 179, 189
  Potassium bromate 80
  Potassium carbonate 97, 142
  Potassium chlorate 35, 38, 79–80, 98
  Potassium chloride 80, 97
  Potassium nitrate 12, 34–36, 48, 53, 64–68, 74, 76, 97
  Potassium perchlorate 47, 81
  Potassium permanganate 143–144, 187, 192
  Potassium sulfate 97
  Potassium sulfide 97
Poudre B 37, 67
Precious metals 83, 86–87, 106, 111
Prentiss, Augustin M. 117–119, 121, 123–124, 127–129, 136, 144–145
Priestley, Joseph 207
Primary explosives 30, 34, 55, 63, 79, 81
Priming device 31, 168
Princess Mary gift tins 106
Princip, Gavrico 54
Procaine, see Novocaine
Propaganda 110–111
Propellants 31–32, 36–38, 43–44, 48–49, 56, 60, 63–68, 70–71, 75–76, 79, 81, 96
Protozoa 174–175, 185
Puddling 88–89
Pyrolusite, see manganese dioxide
Pyrotechnic mixtures 48–49, 98
Pyroxylin, see Guncotton
Quick-fire (QF) weapons 32, 96
Quickmatch 49, 67
Quinine 183–185, 187, 189, 195
Radioactivity 113
Radiography 21, 111–114
Ramsay, William 121–123
Red Star gas 126
Regimental aid posts 159, 161–62, 171
Relapsing fever 182–183
Remarque, Erich Maria 103, 106, 169
Respirators 138–144, 146, 148, 200
Richards, Frank 51, 54, 172, 186
Rifle grenades 55–56, 69, 122
Rogers, Leonard 177
Roland grenade 52
Röntgen, Wilhelm Konrad 21, 112
Royal Army Medical Corps (RAMC) 190

Royal Victoria Hospital, Netley 110, 170–171, 204
Russian Commission for the Production of Asphyxiating Gases (CPAG) 130–131, 146
Russo–Japanese War 54
Rusting 86, 93
Saline 162–164, 177, 186, 197
Saltpetre, see Potassium nitrate
Salvarsan 186–187
Sanitation 114, 166, 170–171, 190–191, 212, 214
Sassoon, Siegfried 51, 54, 192, 202
Scheele, Car Wilhelm 21, 125
Scheer, Admiral Reinhard 45
Schneiderite 78
Schönbein, Christian Friedrich 67–68
Schulze, Johann 107
Scopolamine 162, 207
Second Industrial Revolution 87–88, 96, 102
Serum 100, 180, 184, 190, 195
Sexually transmitted infections 165, 186–187, 190
Shell shortage 29, 42, 126
Shellac 54, 98
Shells, see Artillery shells
Shrapnel shells 29–30, 32, 40, 42, 46, 60, 65, 100, 126, 179
Shrapnel, Henry 40
Sibert, Major General William L. 148–149
Siege artillery 31
Siege of Kut 168
Siemens-Martin process 90–92
Signal rockets 49, 97–98
Silicate minerals 102
Silicon 33, 88–89, 92, 102
Silver 20, 83, 86, 95–96, 106–109, 111–113
Silver compounds 107–109, 111–112, 187, 194
Single-based propellant 70
SK gas 123
Slag 88–90, 93–94
Slow match 53–54, 67
Small Box Respirator 143–144
Smelting 12, 86, 88, 93–94, 96
Smoke blanket 45
Smoke bombs 44
Smoke screen 45
Smokeless powder 37, 60, 63, 65, 67–68, 70–71, 75–76, 81, 152
Smoke-producing materials 37, 46, 57
Snout-type canister mask 138, 142–143
Sobrero, Ascanio 69
Soda lime 143–144, 149
Sodium 48, 106, 177
Sodium compounds 80, 97–98, 135, 139, 143, 150, 156, 176, 196
  Sodium bicarbonate 139, 145
  Sodium carbonate 139, 141, 150, 158, 181, 196
  Sodium chloride 97, 107, 125, 162, 177
  Sodium hydrogensulfate 192
  Sodium hydrosulfite 156
  Sodium hydroxide 48, 102, 125, 140–141, 143, 149–150, 158, 196
  Sodium hypochlorite 125, 194–196
  Sodium iodide 133
  Sodium nitrate 16, 23, 25, 65, 84–85
  Sodium permanganate 144, 149–150
  Sodium phenate 140–141
  Sodium ricinoleate 141
  Sodium sulfanilate 141
  Sodium thiosulfate 108, 139, 141, 145
Solder 95, 98, 106
Spanish flu 188–189, 203
Spanish–American War 153
Spring Offensive 1918 124, 179
Stahlhelm 92
Stannic chloride 57, 126, 131
Star shells 33, 48–49, 67, 97–98, 100
Steel 17–18, 21, 28, 33–35, 38–41, 43, 47, 53, 56–57, 86, 89–98, 103, 105–106, 123,
213
Steel helmets 92–93, 213
Steiner, Frederick 156–157
Stenhouse, John 129–130, 138
Sternutators 42–43, 118–119, 123–124, 126
Stick grenades 55–56, 69, 106
Stokes mortar 43, 81, 123
Stoney, Florence 113–114
Stovocaine 208
Strontium carbonate 49, 98
Strontium nitrate 48
Strychnine 183, 189
Submarines 21, 73, 95, 102, 164, 211
Sulfonamides 187
Sulfonation 74
Sulfur 12, 23, 36–37, 46, 64–66, 79, 83, 93–94, 97, 117, 134, 175
  Sulfur black dye 156
  Sulfur brown dye 156
  Sulfur dichloride 126
  Sulfur dioxide 83–84, 193
  Sulfur mustard 135–136
  Sulfur trioxide 45–46, 83–84, 128
Sulfur-free gunpowder 37, 65
Sulfuric acid 16, 22, 45, 56, 67, 69, 74–75, 80, 83–86, 94, 128, 131, 150, 157, 204–205
Synthetic organic chemicals 13–14, 23–24, 76, 84, 88, 125, 127, 152, 154–157, 186–
187, 197, 208, 213
Syphilis 186–187
T shells 41
Tanks 19, 20–21, 91–92, 213
Tannins 154
Tear gas, see Lachrymatory agents
Tetanus 179–180
Tetanus antitoxin 180, 184, 190, 203
Tetryl 35–36, 43, 80–81
Thermite 47–48, 57, 81, 100–101
Tin 103–106
  Tin cans 103–106, 133, 213
  Tin dioxide 96
  Tin oxide 46
Tincture of iodine 163, 179, 194
Tincture of opium, see Laudanum
Tinstone, see Tin dioxide
Tissot mask 143
Titanium tetrachloride 46
TNT poisoning 62–63
TNT 13, 15, 24–25, 30, 34, 39, 41–43, 60–63, 74–78, 155
Toluene 75–77
Tonite 56, 68
Torpedoes 21, 68, 164–165
Tracer shells 48, 106
Trench foot 184
Trench nephritis 180–181
Trench warfare 15, 52, 54, 126, 154
Trinitrotoluene, see TNT
Trotyl, see TNT
T-Stoff 41, 122, 126
Tube-helmet, see British Phenate Helmet
Tungsten 17, 21, 93–94
Tunnel warfare 20, 64, 66, 68, 75, 77–78, 117, 211, 214
Turkey red, see Alizarin
Turkey red oil 150
Turpin, François Eugène 15, 74
Typhoid 175–177, 182, 190–191, 197
Typhus 172, 181–183
Uranium 113
Urine 62, 64, 138, 176, 181, 187
Urotropin, see Hexamine
Vaccines 188–191
Vanadium pentoxide 84
Venereal diseases, see Sexually transmitted infections
Vent sealing tube 37, 68
Verdigris 86
Vermin 172, 192–193
Vermorel sprayers 145
Veronal 203
Very flares 49
Vesicants 41–42, 118–119, 131, 134–136, 144–145, 214
Vieille, Paul 67
Vincennite 131, 134
Viruses 173–174, 181–182, 188
Voluntary Aid Detachment (VAD) 189
Water purification 125, 181, 191–192, 213
Water supplies 172, 191–192
Webley revolver 71
Weizmann, Chaim 73
White Cross shells 119
White Star gas 116, 126
Whiteley, Martha 209
Wilbrand, Julius 75
Women's National League for the Conservation of Platinum 87
Wright, Sir Almroth 190, 197
Wrought iron 88–90, 97–98, 105
Xhosa War 153
X-rays 21, 111–114, 162
Xylyl bromide 41, 56, 118, 122, 126
Yellow Cross shells 42, 46, 119, 135
Yellow Star gas 126
Yperite 135
Zeebrugge raid 70
Zelinskii-Kummant mask 143, 147
Zeppelin 101–102
Zinc 23, 93–96, 98–99, 143
  Zinc oxide 143–144, 193